Seventh-Grade Math Minutes

One Hundred Minutes to Better Basic Skills

Written by
Doug Stoffel

Editor:	Sue Jackson
Senior Editor:	Maria Elvira Gallardo, MA
Cover Illustrator:	Rick Grayson
Production:	Libby Kraten, Sandra Riley
Cover Designer:	Barbara Peterson
Art Director:	Moonhee Pak
Managing Editor:	Betsy Morris, PhD

TABLE OF CONTENTS

Introduction

Seventh grade is an extremely important math year in the lives of students. It is often one of the final years for students to solidify their basic math skills before moving on to the abstract world of algebra and geometry. The focus of *Seventh-Grade Math Minutes* is math fluency—teaching students to solve problems effortlessly and rapidly. The problems in this book provide students with practice in every key area of seventh-grade math instruction, including:

- computation
- number sense
- graphing
- problem solving
- measurement
- data analysis and probability
- spatial connections
- reasoning and proof
- algebra and functions
- communication
- geometry

Use this comprehensive resource to improve your students' overall math fluency, which will promote greater self-confidence in their math skills as well as provide the everyday practice necessary to succeed in testing situations.

Seventh-Grade Math Minutes features 100 "Minutes." Each Minute consists of 10 classroom-tested problems of varying degrees of difficulty for students to complete within a one- to two-minute period. This unique format offers students an ongoing opportunity to improve their own fluency in a manageable, nonthreatening format. The quick, timed format, combined with instant feedback, makes this a challenging and motivational assignment students will look forward to using each day. Students become active learners as they discover mathematical relationships and apply acquired understanding to complex situations and to the solution of realistic problems in each Minute.

How to Use This Book

Seventh-Grade Math Minutes is designed to be implemented in numerical order, starting with Minute One. Students who need the most support will find the order in which skills are introduced most helpful in building and retaining confidence and success. For example, the first time that students are asked to provide the value of pi to the hundredths place, the digits in the ones and tenths place are provided. The second time, the digit in the ones place is provided. It is not until the third time that students are asked the value of pi that they must recall the number without additional support.

Seventh-Grade Math Minutes can be used in a variety of ways. Use one Minute a day as a warm-up activity, bell work, review, assessment, or a homework assignment. Other uses include incentive projects and extra credit. Keep in mind that students will get the most benefit from their daily Minute if they receive immediate feedback. If you assign the Minute as homework, correct it in class as soon as students are settled at the beginning of the day.

If you use the Minute as a timed activity, place the paper facedown on the students' desks or display it as a transparency. Use a clock or kitchen timer to measure one minute—or more if needed. As the Minutes become more advanced, use your discretion on extending the time frame to several minutes if needed. Encourage students to concentrate on completing each problem successfully and not to dwell on problems they cannot complete. At the end of the allotted time, have the students stop working. Then read the answers from the answer key (pages 108–112) or display them on a transparency. Have students correct their own work and record their scores on the Minute Journal reproducible (page 6). Then have the class go over each problem together to discuss the solution(s). Spend more time on problems that were clearly challenging for most of the class. Tell students that problems that seemed difficult for them will appear again on future Minutes and that they will have another opportunity for success.

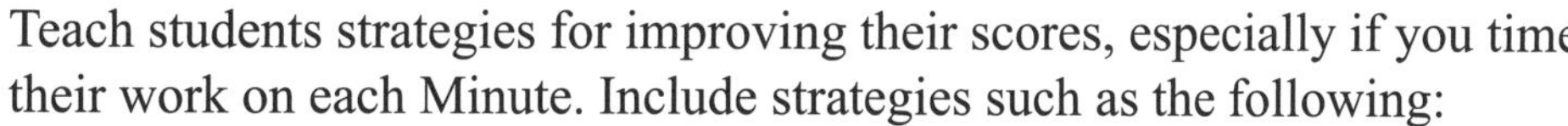

Teach students strategies for improving their scores, especially if you time their work on each Minute. Include strategies such as the following:

- leave more time-consuming problems for last
- come back to problems they are unsure of after they have completed all other problems
- make educated guesses when they encounter problems with which they are unfamiliar
- rewrite word problems as number problems
- use mental math whenever possible
- underline important information
- draw pictures

Students will ultimately learn to apply these strategies to other timed-test situations.

The Minutes are designed to improve math fluency and should not be included as part of a student's overall math grade. However, the Minutes provide an excellent opportunity for you to see which skills the class as a whole needs to practice or review. This information will help you plan the content of future math lessons. A class that consistently has difficulty reading graphs, for example, may make excellent use of your lesson in that area, especially if the students know they will have another opportunity to achieve success in reading graphs on a future Minute. Have students file their Math Journal and Minutes for the week in a location accessible to you both. You will find that math skills that require review will be revealed during class discussions of each Minute. You may find it useful to review the week's Minutes again at the end of the week with the class before sending them home with students.

While you will not include student Minute scores in your formal grading, you may wish to recognize improvements by awarding additional privileges or offering a reward if the entire class scores above a certain level for a week or more. Showing students that you recognize their efforts provides additional motivation to succeed.

Minute Journal

Name ___________________________

Minute	Date	Score	Minute	Date	Score	Minute	Date	Score	Minute	Date	Score
1			26			51			76		
2			27			52			77		
3			28			53			78		
4			29			54			79		
5			30			55			80		
6			31			56			81		
7			32			57			82		
8			33			58			83		
9			34			59			84		
10			35			60			85		
11			36			61			86		
12			37			62			87		
13			38			63			88		
14			39			64			89		
15			40			65			90		
16			41			66			91		
17			42			67			92		
18			43			68			93		
19			44			69			94		
20			45			70			95		
21			46			71			96		
22			47			72			97		
23			48			73			98		
24			49			74			99		
25			50			75			100		

SCOPE AND SEQUENCE

SKILL	MINUTE IN WHICH SKILL FIRST APPEARS
Order of Operations	1
Whole Numbers (add, subtract, multiply, divide)	1
Fractions (add, subtract, multiply, divide, equivalent, reducing)	1
Perimeter	1
Graphs (Bar, Line, Circle)	1
One-step Algebra Equations	1
Patterns/Sequences	1
Algebraic Substitution/Expressions	2
Area (squares, rectangles, parallelograms)	2
Exponents/Squares/Square roots	2
Money	2
Bar Notation	3
Inequalities	3
Spatial Reasoning	3
Multiplying and Dividing by 10 and Powers of 10	4
Decimals (addition, subtraction, multiplication, division)	4
Estimation	4
Percentages	4
Nets	4
Coordinate Graphs (rows and columns)	4
Problem Solving/Applied Math	5
Venn Diagrams	6
Geometry (congruent, similar, shapes, vertices, sides, degrees, vocabulary)	7
Place Value	8
Number Sense and Reasonable Answers	8

SKILL	MINUTE IN WHICH SKILL FIRST APPEARS
Factors/Multiples	9
Probability	10
Symmetry	10
Integers (add, subtract, multiply, divide)	12
Prime and Composite Numbers	12
Ratios	14
Divisibility	15
Time	15
Number Lines	19
Ordering and Comparing Numbers and Amounts	22
Circles (diameters, radius)	23
Analogies	25
Like Amounts	30
Frequency Tables	41
Volume	51
Function Rules	52
Coordinate Grids	53
Lines (parallel, perpendicular, intersecting, slopes, intercepts)	53
Angles (right, obtuse, acute)	60
Surface Area	61
Stem-Leaf Plots	71
Math Crossword Puzzles	72
Mean/Median/Mode	74
Percent Increase and Decrease	76
Absolute Value	89
Recognizing Errors	91

MINUTE 1

1. Simplify: $12(2 + 7 + 1) =$ _10_

2. $\dfrac{3}{10} \cdot \dfrac{7}{10} =$ _$\frac{1}{10}$_

3. Circle all of the following equal to $\dfrac{2}{5}$:   0.4 $\dfrac{4}{100}$ 40% 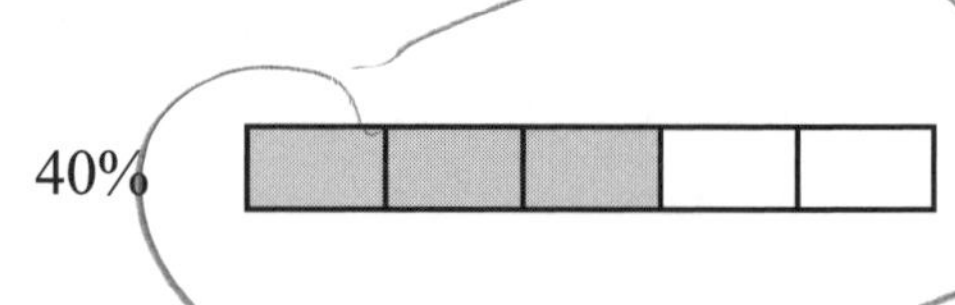

4. $10 \cdot \boxed{2} = 5$

5. Cross out the three-dimensional shape. 

6. Each side of the regular pentagon is 5 centimeters. What is the perimeter? _15_ 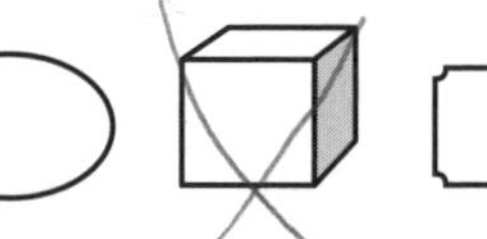

7. In the graph, Alex has ___5___ times as much money as Annie.

8. If $a = 5$ and $b = 4$, then $2a + b =$ _14_ .

9. If $3x = 27$, then $x =$ _3_ .

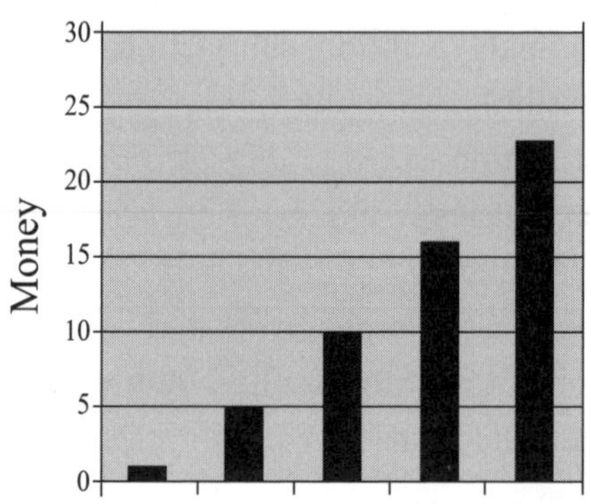

10. Which of the following shapes comes next in the pattern? 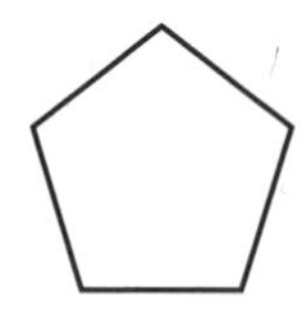

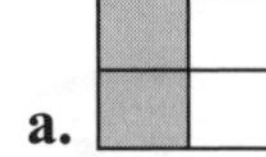 **a.**   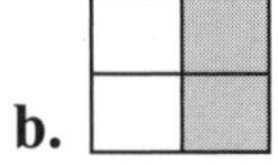 **b.**   **c.**   **d.**

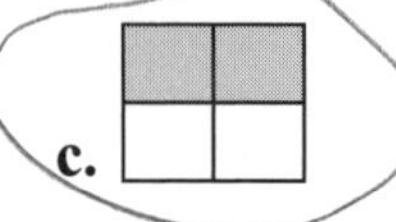

MINUTE 2

1. $\dfrac{12}{2} \cdot \dfrac{1}{3} =$

2. Use the correct symbol (=, >, or <) to complete: $\dfrac{3}{10} + \dfrac{7}{10}$ ☐ $\dfrac{3}{10} \cdot \dfrac{7}{10}$

3. Which of the following does not belong? Circle your answer.

Two-tenths   0.2   20%   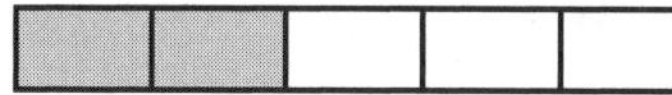

4. The distance between two cities would most likely be measured in:
a. feet   **b.** inches   **c.** yards   **d.** miles

5. The shaded area in figure B is ______ times greater than the shaded area in figure A.

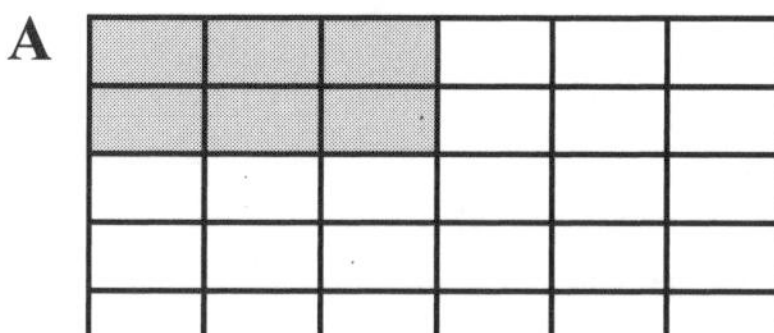 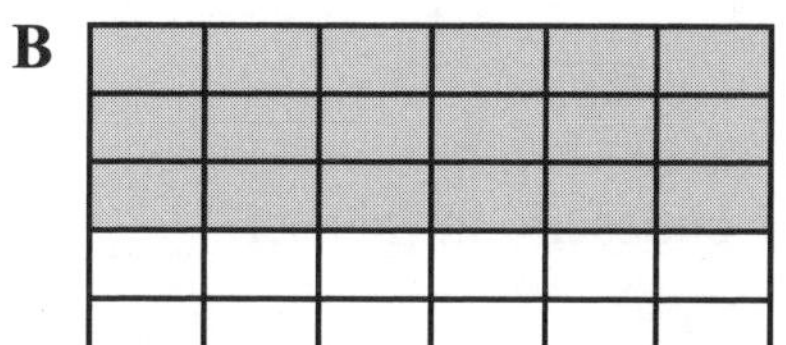

6. The perimeter around the shaded area in figure A in Problem 5 is _______ units.

7. In the graph, _______________ has five times as much money as _______________.

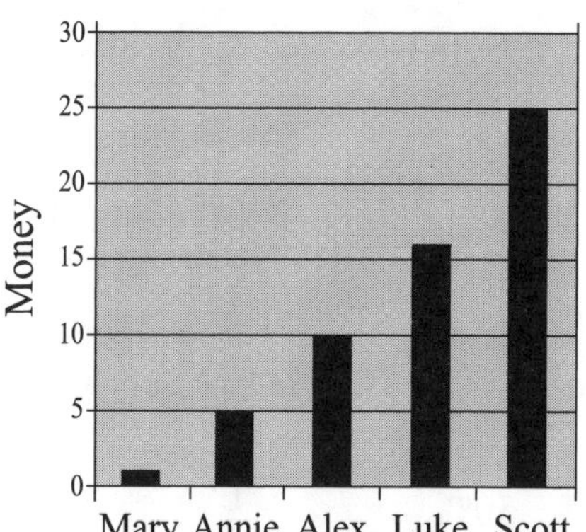

For Problems 8–10, evaluate if $a = 4$, $b = 6$, and $c = 2$.

8. $ab =$

9. $\dfrac{a + b}{c} =$

10. $b^2 =$

MINUTE 3

1. $2\left[\dfrac{30}{5}\right] =$

2. $\left(\dfrac{1}{4}\right)\left(\dfrac{1}{3}\right) =$

3. Which of these represents the greatest amount?

Circle: 62% $\dfrac{1}{2}$ 0.58   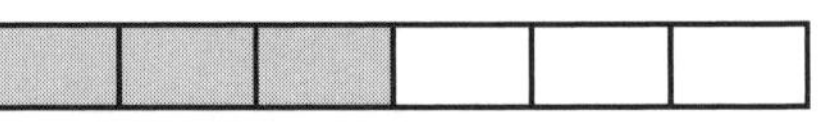

4. Use •, +, −, or ÷ to complete the following equation.   2 ☐ 4 ☐ 1 = 9

5. How many cubes are in this set? _______ 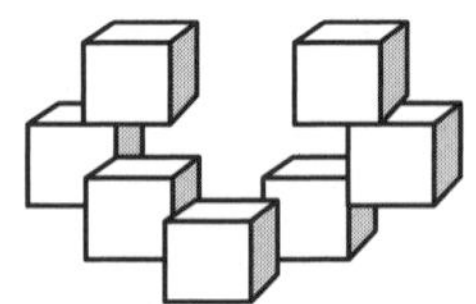

6. The distance around the world at the equator is about 42,000 ____________.
   a. meters   **b.** kilometers   **c.** centimeters   **d.** millimeters

7. What number will complete the box? _______ 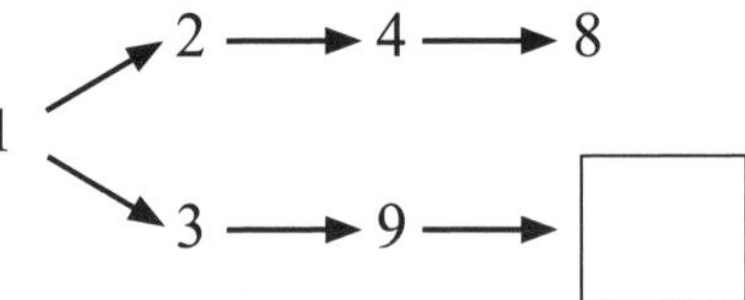

For Problems 8–10, use >, <, or =.

8. 50% _______ $\dfrac{1}{2}$

9. 3^2 _________ 2^3

10. $0.\overline{5}$ _______ 0.5

MINUTE 4

1. $0.7 \times 8 =$

2. $576 \div 10 =$

3. If $\dfrac{2}{5} + \dfrac{x}{5} = \dfrac{7}{5}$, then $x =$ _______________.

4. If $\left[\dfrac{3}{8}\right] \cdot \left[\dfrac{a}{2}\right] = \dfrac{15}{16}$, then $a =$ _______________.

5. In the graph, shade column A and put an X in E4.

6. What shape would the net to the right create if you folded it?

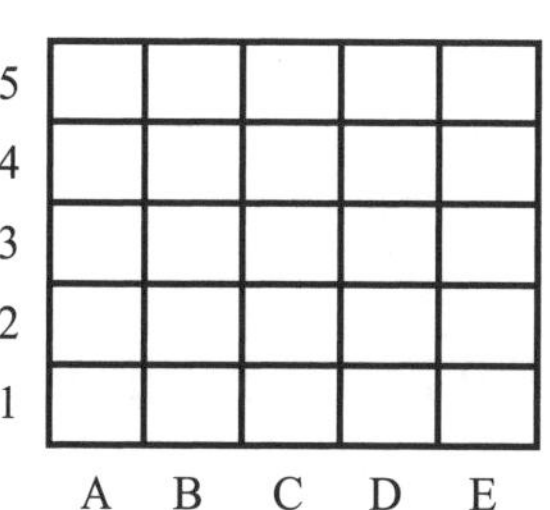

a.   **b.**

c.   **d.**

7. About what percent of the graph does region A represent?
a. 50% **b.** 90% **c.** 10% **d.** 33%

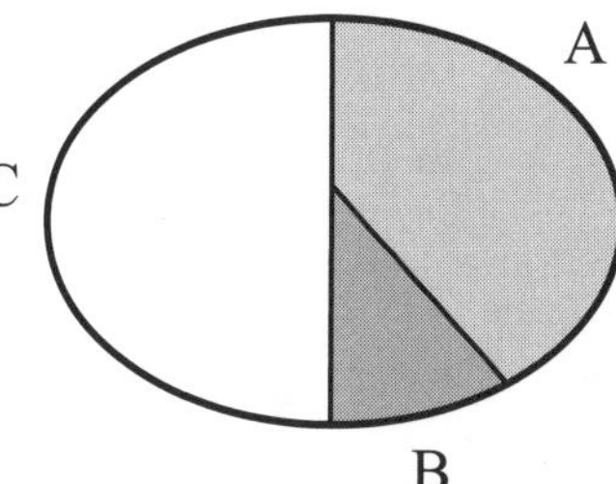

For Problems 8–10, estimate to find the best answer.

8. 19 out of 80:
a. 10% **b.** 40% **c.** 25% **d.** 75%

9. 9% of 55:
a. 50 **b.** 30 **c.** 20 **d.** 5

10. 194% of 40:
a. 225 **b.** 75 **c.** 40 **d.** 30

MINUTE 5

1. $0.5 \times 0.9 =$

2. $3 + 2 \cdot 4 + 5 =$

3. Which of these represents the least amount?

Circle: 0.35 $\dfrac{12}{50}$ 25%

4. Fill in the remaining prime numbers that are less than 20.

| 2 | | 7 | | 13 | | |

5. Shade row 3 and column C.

6. At what point does the row and column shaded in Problem 5 intersect? _______

7. In 1933, Wiley Post flew around the world in 7 days, 18 hours. Wiley's trip would best be described as flying around the _______ of the earth.
a. perimeter   **b.** area   **c.** volume   **d.** diameter

8. Find the number that completes the following problem.

$$\begin{array}{r} 2\square \\ \times\,8 \\ \hline 192 \end{array}$$

9. Find the number that completes the following problem.
$(3 + 5) + 2 = 2(\square + 2)$

10. If $3 \times 3 \times 3 \times 3 = 3^x$, then $x =$ _______.

MINUTE 6

1. $0.3 + 0.5 + 0.8 =$

2. $(2 + 0.4 + 0.6)^2 =$

3. Fill in the remaining positive factors of 18.

1		3	6		18

For Problems 4–6, use the Venn diagram to the right.

4. _______ people liked vanilla only.

5. _______ people liked chocolate only.

6. _______ people liked both.

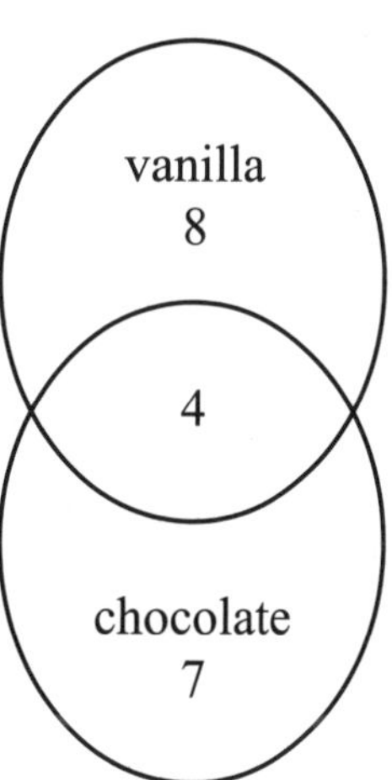

For Problems 7–10, circle *True* or *False*.

7. $\dfrac{8}{8} > \dfrac{12}{12}$ True   or   False

8. $\dfrac{12}{50} = \dfrac{6}{25}$ True   or   False

9. $2.2 > 2.0\overline{9}$ True   or   False

10. $8.15 = 8 + \dfrac{1}{10} + \dfrac{5}{100}$ True   or   False

MINUTE 7

1. $(0.6)^2 =$

2. If $\left[\dfrac{2}{5}\right]^2 = \left[\dfrac{x}{25}\right]$, then $x =$ _______.

3. Circle the greatest number. Cross out the least number.

   $\dfrac{78}{100}$ $\dfrac{3}{4}$    50%

4. Circle the numbers that are multiples of 7.

   21   14   1   17   35

5. Circle the figure that is congruent to  .

   a.    b.   c.    d.

6. What is the perimeter of this figure? _______   10 cm   8 cm

7. Is the area of the figure in Problem 6 greater than or less than 80 cm^2? ___________

8. Find the number that completes the following problem. $42\boxed{} \times 6 = 2{,}538$

9. If $y = x + 5$ and $x = 3$, then $y =$ _______.

10. If $y = x + 5$ and $y = 11$, then $x =$ _______.

MINUTE 8

1. Circle all of the following that are between 10 and 40.

3^2 4^2 5^2 6^2 7^2

2. What is the value of the underlined digit in the number 328.0$\underline{6}$?

a. $\dfrac{6}{10}$ **b.** $\dfrac{6}{100}$ **c.** $\dfrac{6}{1,000}$ **d.** $\dfrac{6}{10,000}$

3. $\left[\dfrac{1}{2}\right]\left[\dfrac{2}{3}\right]\left[\dfrac{3}{4}\right] =$

4. Circle the fractions that reduce to $\dfrac{1}{4}$: $\dfrac{2}{8}$ $\dfrac{4}{12}$ $\dfrac{3}{12}$ $\dfrac{12}{38}$

5. In about how many seconds could a 9-year-old boy run 100 meters?
a. 5 sec.   **b.** 10 sec.   **c.** 20 sec.

6. How many cubes are shown? _______ 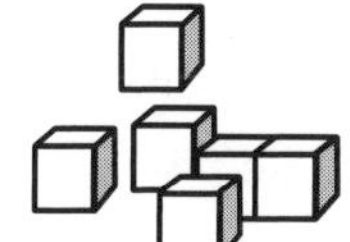

7. Based on this graph, is Mark's company doing well? _______

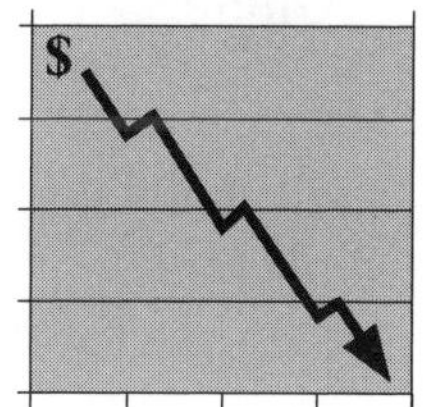

8. Look for the pattern between rows A and B and complete the grid.

A	2	5	7	12
B	5	8	10	

For Problems 9–10, evaluate if $a = 5$, $b = 3$, and $c = 2$.

9. $2ab =$

10. $\left[\dfrac{6}{b}\right]^c =$

MINUTE 9

1. Use the numbers 3, 4, and 5 to complete the math sentence.

$\square + \square \cdot \square = 19$

2. Find the next number in the following sequence: $\dfrac{1}{12}$, $\dfrac{3}{12}$, $\dfrac{5}{12}$, _______.

3. What is 10% of 300? _______

4. How many minutes are in 3 hours and 10 minutes? _______

For Problems 5–7, use the graph to the right.

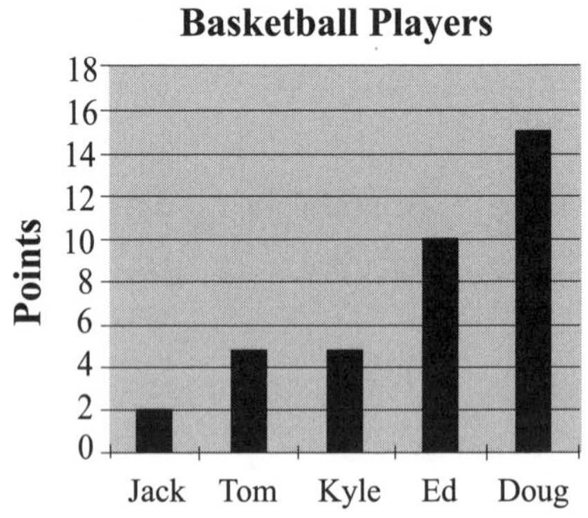

5. Which two players scored the same number of points? _______

6. Ed scored twice as many points as Tom.
Circle: True   or   False

7. How many total points were scored by the players? _______

8. Annie puts $10 into a vacation jar each week. How much will she have saved by the end of the year? _______

For Problems 9–10, use the diagram to the right.

9. Draw arrows to connect the multiples between circles A and B.

10. Circle the numbers in the diagrams that are evenly divisible by 4.

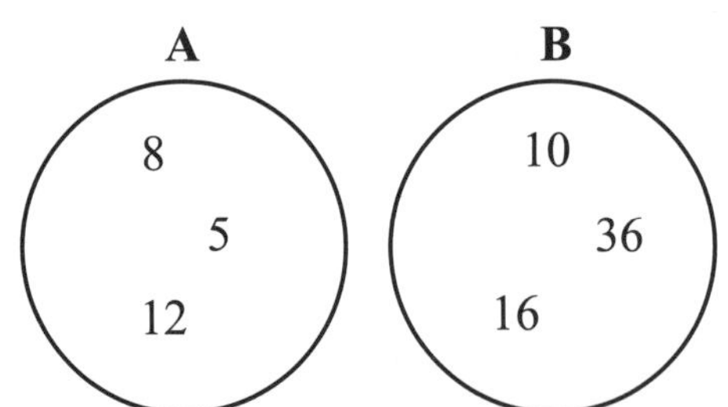

MINUTE 10

For Problems 1–3, circle *True* or *False*.

1. $2 \times 6 \times 3 \times 0 \times 4 > 12 \times 1 \times 1$ ⠀⠀⠀⠀ True ⠀or⠀ False

2. $\sqrt{16} = 4$ ⠀⠀⠀⠀ True ⠀or⠀ False

3. $2^3 = 6$ ⠀⠀⠀⠀ True ⠀or⠀ False

4. Circle each of the following that are whole numbers: ⠀ $\dfrac{12}{2}$ ⠀ $\dfrac{2}{12}$ ⠀ $\dfrac{8}{8}$ ⠀ 2^2 ⠀ $\left[\dfrac{1}{2}\right]^2$

5. What is $\dfrac{1}{2}$ of $\dfrac{3}{4}$? _______

6. Draw the line of symmetry on the figure to the right.

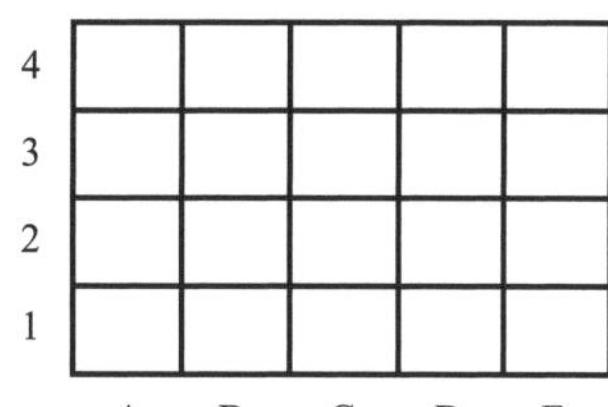

7. Maps often show north as pointing toward the top of the page.
If you went from A2 to E3, in which direction would you be going?
a. NE
b. NW
c. SE
d. SW

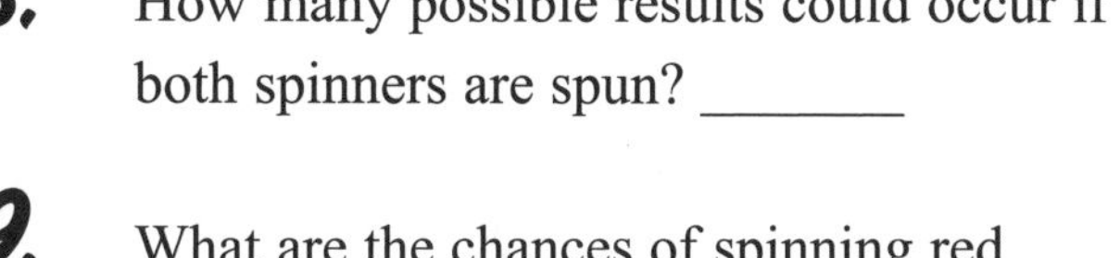

For Problems 8–10, use the spinners to the right.

8. How many possible results could occur if both spinners are spun? _______

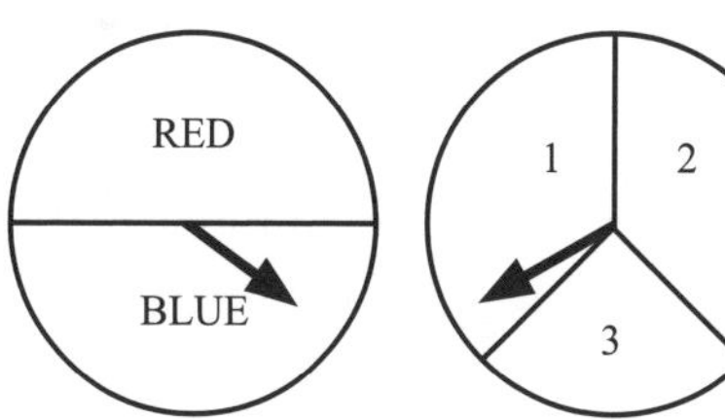

9. What are the chances of spinning red and 3? _______

10. What are the chances of spinning blue and an odd number? _______

MINUTE 11

1. Complete the following factor tree.

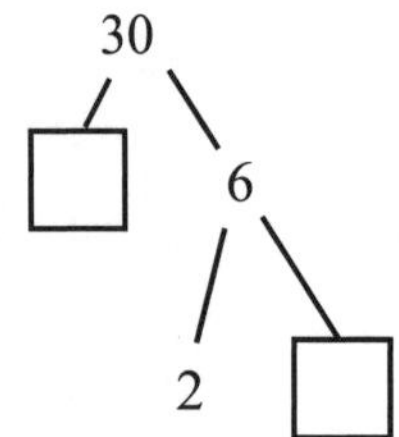

2. $3(4 + 6) - 10 =$

For Problems 3–4, use the table to the right.

3. Which square does not touch one of the perimeter squares? _______

4. What is the combined area of rows 4 and 5? _______

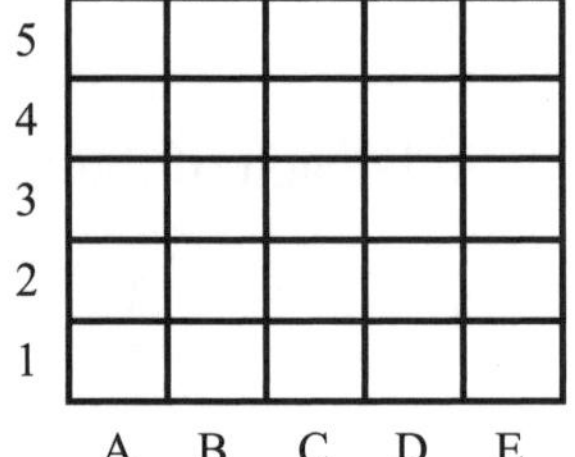

For Problems 5–8, round to the underlined digit. (Note: "≈" means "approximately")

5. $2\underline{7}.38 \approx$ _______

6. $\underline{2}.99 \approx$ _______

7. $3.1\underline{6}7 \approx$ _______

8. $1,001.\underline{4}5 \approx$ _______

For Problems 9–10, use $a = 10$ and $b = 2$.

9. The product of a and b is _______.

10. Three more than twice b is _______.

MINUTE 12

1. $\dfrac{5}{4} - \dfrac{1}{2} =$

2. If $\dfrac{3}{8} \div \dfrac{2}{3} = \dfrac{3}{8} \cdot \dfrac{3}{x}$, then $x =$ _______.

3. $(-4^2) = (-4)(-4)$ Circle: True   or   False

4. $12 \cdot \boxed{} = 4$

5. Which of the following could be the area of a room?
a. 18 m^3 **b.** 50 ft.   **c.** 29 m^2

6. Which answer choice in Problem 5 could be the perimeter of a room? _______

7. Draw two lines in the following trapezoid to create three equilateral triangles.

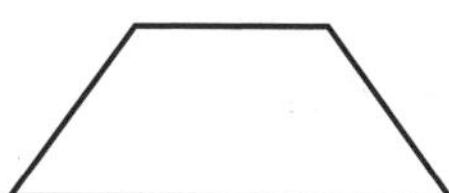

8. What shape would the net to the right create if you folded it?

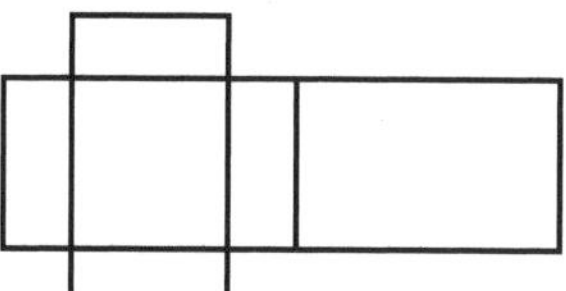

a. 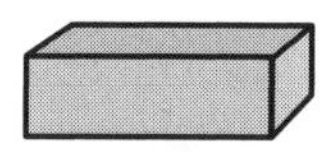   **b.** 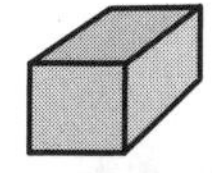

c. 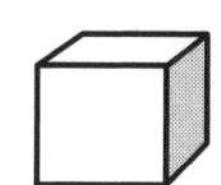   **d.** 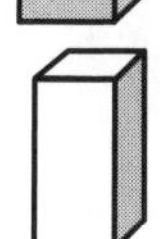

For Problems 9–10, use the Venn diagram to the right.

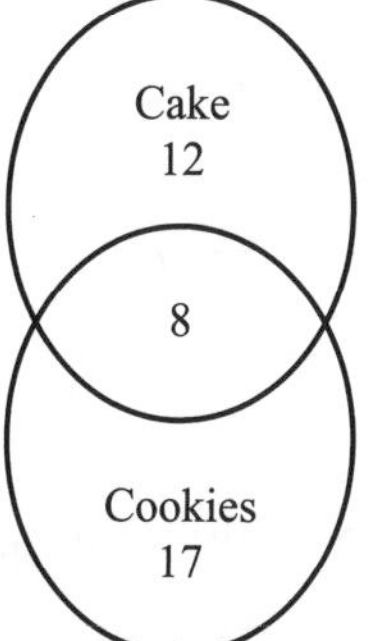

9. How many kids like cookies only? _______

10. How many kids like both cookies and cake? _______

Minute 13

1. $(9 - 3 \cdot 2)^2 =$

2. $205 \times 0.01 =$

3. Rewrite using bar notation: $0.912912... =$ _______

4. Which of the following is the remainder of 14 divided by 3?
a. 4   **b.** 1   **c.** 5   **d.** 2

5. Fill in the remaining prime numbers between 20 and 50.

23	29			41		47

For Problems 6–7, use the graph to the right.

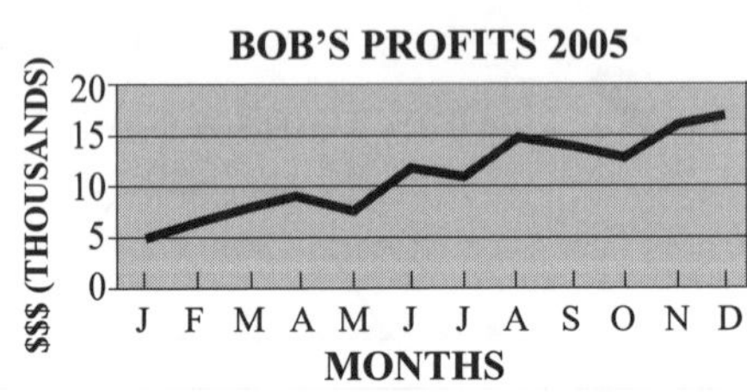

6. Would it be a good idea to invest in Bob's company?
Circle: Yes   or   No

7. In the graph, what does the "F" stand for?
____

For Problems 8–10, estimate to find the best answer.

8. 24 out of 99:
a. 10%   **b.** 75%   **c.** 25%   **d.** 50%

9. 12% of 400:
a. 15   **b.** 40   **c.** 60   **d.** 80

10. Possible weight of a 7th grader:
a. 50 kilograms   **b.** 50 grams   **c.** 50 milligrams

MINUTE 14

1. If $24 = 3 \cdot 2^x$, then $x =$ _______.

2. If $\dfrac{3}{5} = \dfrac{x}{15}$, then $x =$ _______.

3. Find the remaining multiples of 7 that are less than 50.

7		21	28			49

4. Complete the factor tree.

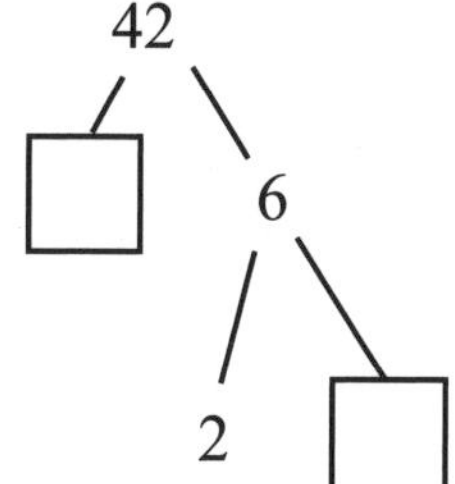

5. Use the digits 5, 7, and 2 to write four numbers that are greater than 400.

_____________ _____________ _____________ _____________

For Problems 6–10, match each math expression with its equivalent expression.

6. $a \div 2$ **a.** $a \cdot a$

7. $a \cdot 2$ **b.** $3a$

8. a^2 **c.** 0

9. $a + a + a$ **d.** $\dfrac{a}{2}$

10. $0a$ **e.** $2a$

MINUTE 15

1. $\dfrac{6}{0.5} =$

2. What is the remainder of 21 divided by 4? ________

3. Is $\sqrt{47}$ closer to 6 or 7? ________

4. Place ( ) symbols in this problem to make a true statement: $4 + 5 \cdot 2 = 18$

5. $1.435 \times 10^2 = 143.5$ Circle: True   or   False

6. If $5.48 = 5 + \dfrac{a}{10} + \dfrac{8}{b}$, then $a =$ ________ and $b =$ ________.

7. Half of a circle is a ________.
   a. square   **b.** triangle   **c.** diamond   **d.** semicircle

8. Shade the figure with the fewest vertices. Cross out the figure with the most vertices.

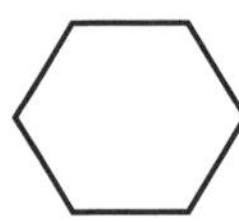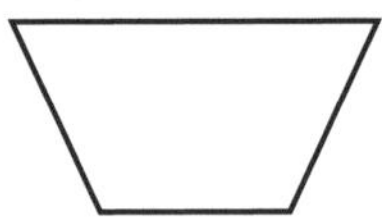

9. If it is 4 o'clock now, what time will it be in 9 hours? ________

10. Which one of the following shapes comes next in the pattern? 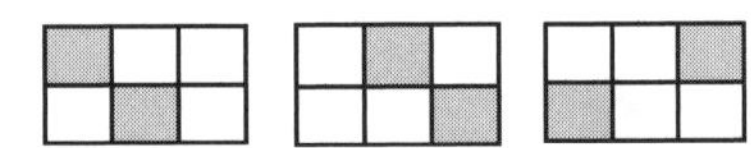

   a. 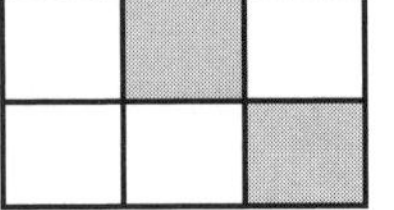   **b.** 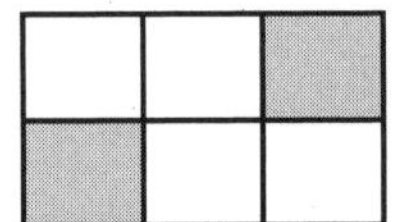   **c.** 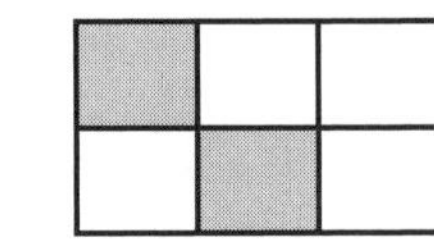

MINUTE 16

1. Circle the greatest number. Cross out the least number.

3.03   3.3   3.003   0.3   0.33

2. Circle the number that is divisible by 4:   45   38   32   30

3. What is the value of the underlined digit in 4$\underline{7}$8.6?

a. 7   **b.** 70   **c.** 700   **d.** 7,000

4. $24 \cdot \boxed{} = 6$

5. Fill in the missing numbers in the table.

Sum	Product	Numbers
7	12	3 and 4
10	16	__ and __

6. Shade the hexagon.

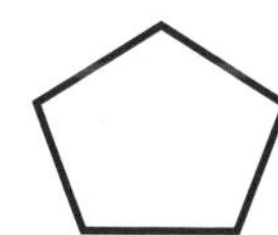   

7. Draw a horizontal line of symmetry through the shape.

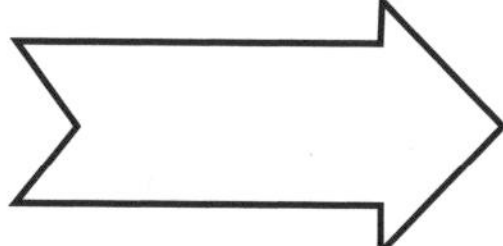

For Problems 8–10, use >, <, or = .

8. $\dfrac{3}{10}$ _______ 0.3

9. $0.\overline{4}$ _______ 0.4

10. 100% of 50 _______ 10% of 600

MINUTE 17

1. In a math problem, which of the following should be done first?
   a. parentheses ( )   **b.** exponents   **c.** multiplication   **d.** addition

2. In a math problem, which of the following should be done last?
   a. parentheses ( )   **b.** exponents   **c.** multiplication   **d.** addition

3. $4\dfrac{1}{4} + 3\dfrac{2}{4} =$

4. $576 \div 10 =$

5. Which of these shapes is congruent to 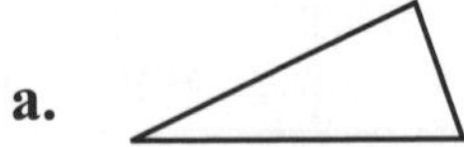 ?

   a.    **b.** 

   c.   **d.** 

For Problems 6–8, use the grid to the right.

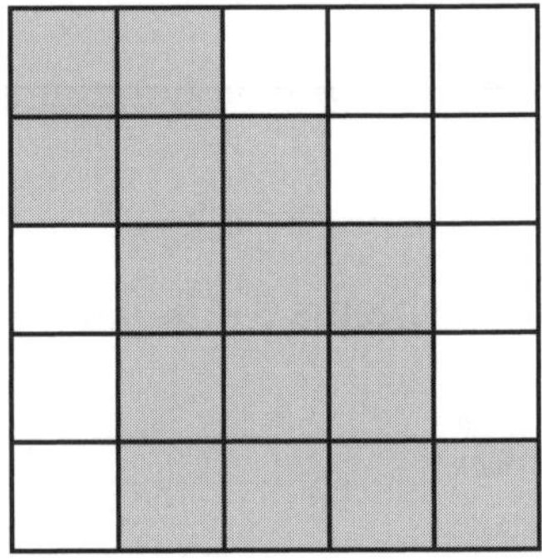

6. What is the area of the shaded region? _______

7. What fraction of the squares in the grid are shaded? _______

8. What percent of the boxes in the grid are shaded? _______

9. If $\dfrac{15}{25} = \dfrac{x}{100}$, then $x =$ _______ .

10. If 60% of a shape is shaded, what percent is NOT shaded? _______

MINUTE 18

1. Fill in the missing fraction: $\dfrac{1}{10}, \dfrac{3}{10}, \dfrac{5}{10}, \boxed{}, \dfrac{9}{10}$

For Problems 2–5, use the graph to the right.

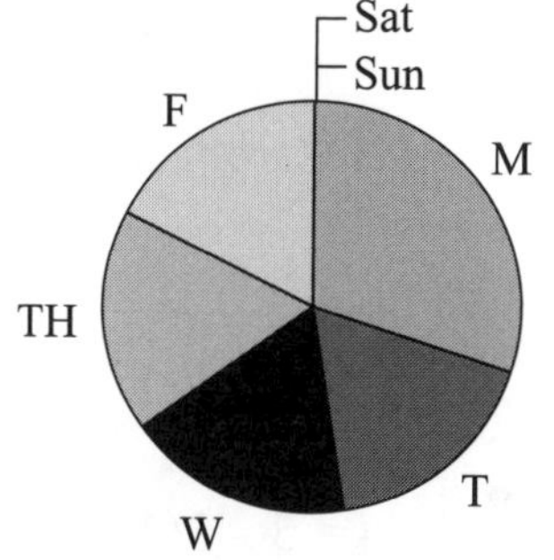

2. On which day of the week did Mark work the most hours? ___________________

3. On which two days of the week does it appear that Mark did not work at all?

___________________ ___________________

4. Is it possible to tell how many total hours Mark worked during this particular week?
Circle: Yes   or   No

5. On Tuesday, Wednesday, and Friday, Mark performed about _______% of his total work for the week.

6. Write the next "A" in this pattern: ◁ A ▷

7. Fill in the missing numbers in the table.

Sum	Product	Numbers
5	6	2 and 3
12	32	__ and __

8. Which of the following does NOT mean a times b?

a. ab **b.** $a \cdot b$ **c.** $a \times b$ **d.** $\dfrac{a}{b}$

9. Which of the following does NOT mean to divide?

a. quotient   **b.** $a \div b$ **c.** ab **d.** $\dfrac{a}{b}$

10. If $\dfrac{1}{5} \div \dfrac{2}{3} = \dfrac{1}{5} \cdot \dfrac{x}{2}$, then $x =$ _______.

MINUTE 19

1. What decimal is the arrow pointing toward? _______

0.9   1.0

2. Round $3.\overline{28}$ to the nearest thousandth. _______

3. If Carol can read 45 pages in one hour, how many pages can she read in four hours?

____

4. $4 \cdot 5 - 3(4) =$

5. Shade 20% of the squares in this box.

6. If you double the sum of 5 and the number _______, you will get 16.

For Problems 7–10, evaluate if $x = 3$, $y = 4$, and $z = 5$.

7. $6(x + y) =$

8. $\dfrac{2}{z - x} =$

9. $2x + 2y =$

10. $\dfrac{1}{2}yz =$

MINUTE 20

1. $18 - 5 \cdot 3 =$

2. $(9 + 4)(10 - 8) =$

3. Is $\sqrt{34}$ closer to 5 or 6? _______

4. If $q - 3.1 = 4.6$, then $q =$ _______.

5. Shade 15% of the box. (**Hint:** 7.5% is already shaded for you.)

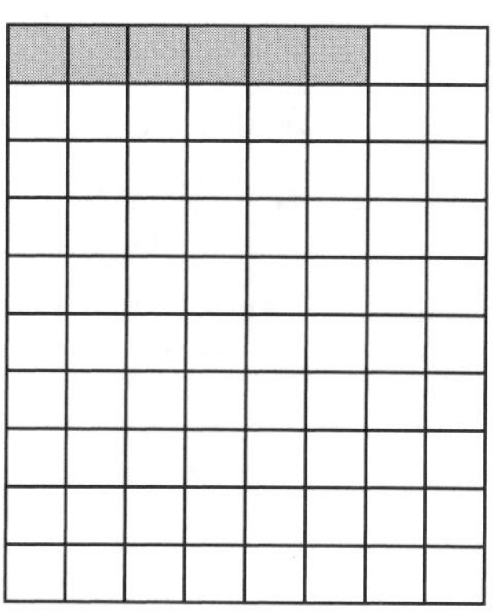

6. Fill in the missing number in the box.

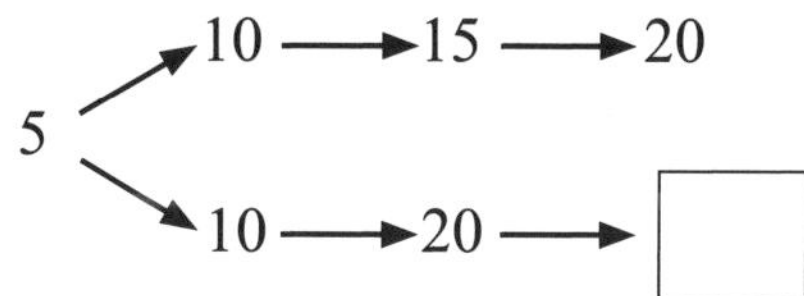

For Problems 7–9, use the Venn diagram to the right.

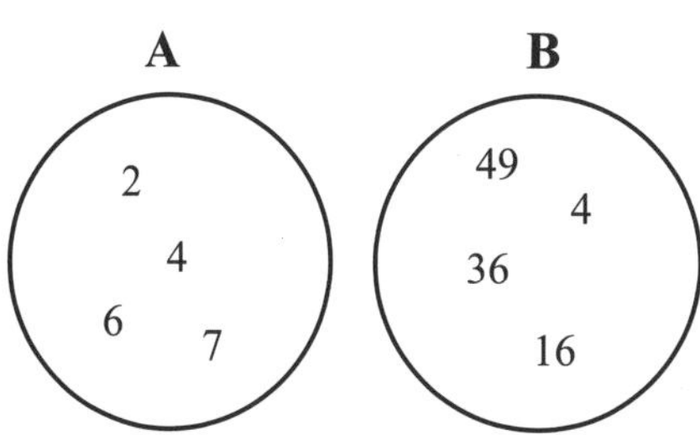

7. Draw arrows to connect the square roots.

8. To which circle would the number 5 belong? _______

9. The sum of the numbers in circle A is a prime number.   Circle: True   or   False

10. If 1 km = 1,000 meters, then $2\frac{1}{2}$ km = _________ meters.

MINUTE 21

For Problems 1–3, circle *True* or *False*.

1. $25 \div 5 \cdot 3 = 15$ True   or   False

2. $2(10 - 7) - 4 = 9$ True   or   False

3. $16 + 24 \div 8 - 5 = 14$ True   or   False

4. Which two grids have the same percentage of squares shaded?

a. 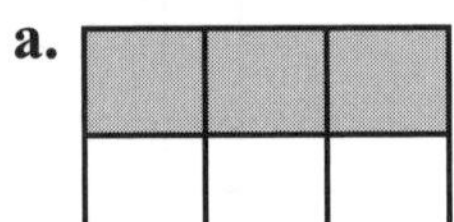   b. 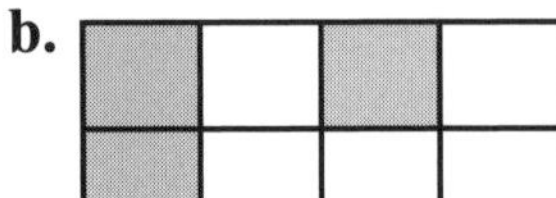   c. 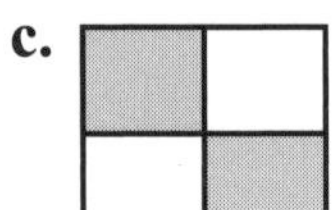   d. 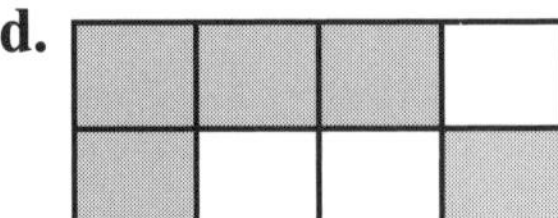

5. Use the numbers 4, 5, and 6 to fill in the circles so that each side equals 11.

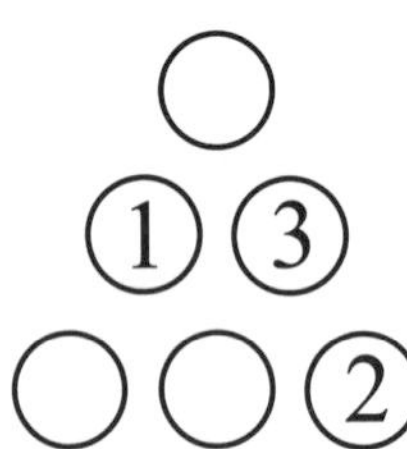

For Problems 6–9, use the graph to the right.

6. How many birthdays were in Jan.–Mar.? _______

7. Were there more boy or girl birthdays in Oct.–Dec.? _______

8. How many girls are in the class? _______

9. How many boys are in the class? _______

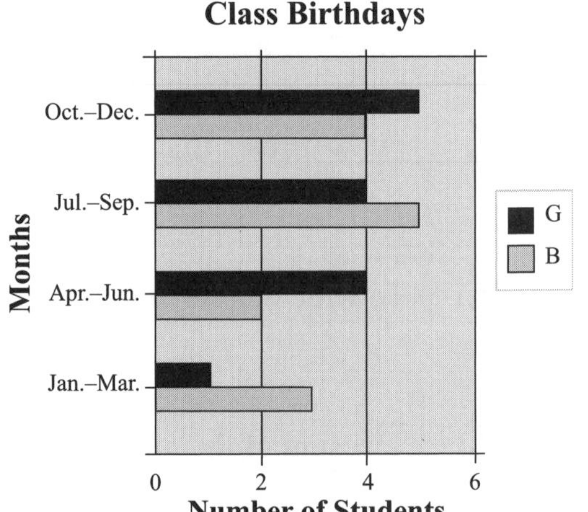

10. Write the next "A" in this pattern: 

MINUTE 22

1. $\dfrac{8}{0.5} =$

2. Which numbers are identified by points A, B, and C on the number line? _______________

3. Order the numbers {10, -7, 8, 0} from least to greatest. _______________

4. $\dfrac{3}{7} \div \dfrac{4}{7} =$

Difference	Product	Numbers
5	6	1 and 6
6	40	__ and __

5. Fill in the missing numbers in the table.

6. Which shape would the net to the right create if you folded it?

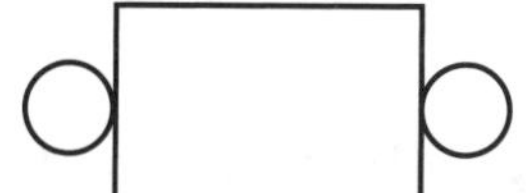

a.

b.

c.

d.

For Problems 7–10, use >, <, or = if $a = 2$, $b = 4$, and $c = 5$.

7. ab _______ ac

8. $b + b$ _______ $2b$

9. $2c - 2b$ _______ 0

10. $2(a + b)$ _______ $2a + 2b$

MINUTE 23

For Problems 1–3, use the grid to the right.

2	4	5	0
1	5	3	9
1	2	9	2
4	7	3	6

1. Circle three consecutive numbers that have a sum of 12.

2. Shade the prime numbers that are greater than 3.

3. Cross out the number that has 2 and 3 as factors.

4. If $\dfrac{d}{7} = 8$, then $d =$ _______.

5. Draw a radius in the circle to the right.

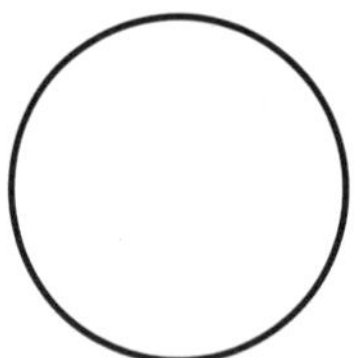

6. If the radius of a circle is 6 cm, the diameter is _______ cm.

7. Draw a vertical line of symmetry on the star.

8. TON is to NOT as 356 is to _______.

   a. 536   **b.** 635   **c.** 635   **d.** 653

9. If you double a number and add 1, you get 11. What is the number? _______

10. If $y = 2x - 4$ and $x = 12$, then $y =$ _______.

MINUTE 24

1. $[1 + (7 - 2)]^2 =$

2. If $a = 3.6$ and $b = 10$, then $ab = $ ________.

3. Write thirty-eight thousandths as a decimal. ________________

For Problems 4–7, use the calendar to the right.

4. What day of the week is March 18? ______________

MARCH

S	M	T	W	T	F	S
				1	2	3
4	5	6	7	8	9	10
11	12	13	14	15	16	17
18	19	20	21	22	23	24
25	26	27	28	29	30	31

5. Circle the date that is three weeks after March 2.

6. Put an "X" on the numbers that are perfect squares.

7. Shade the date that is 15 days before March 26.

8. Round 2,561 to the nearest hundred. ____________

9. 2.5 meters > 220 cm   Circle: True   or   False

10. A coin is tossed three times and lands heads, tails, and tails. The next flip will be:
a. heads   **b.** tails   **c.** unknown

MINUTE 25

1. $10,000 = 10 \times 10 \times \boxed{} \times \boxed{}$

2. If $38,433 = 3.8433 \times 10 m$, then $m =$ ________.

3. $1 + (2)(3)(4) =$

For Problems 4–6, use the grid to the right.

7	9	14	27
2	13	3	28
11	7	15	35
14	18	21	20

4. Shade the multiples of 7.

5. Circle the number in the 2nd row, 2nd column.

6. What is the sum of the numbers in the first column? ________

7. What is the total price of a $5 book with a 10% sales tax? ________

8. If $b^2 = 25$, then $b =$ ________.

9. Circle the expression that shows 15 divided by a number.
  a. $15n$ **b.** $15 - n$ **c.** $15 + n$ **d.** $\dfrac{15}{n}$

10. RAT is to TAR as 246 is to ________.
  a. 624   **b.** 642   **c.** 324   **d.** 236

MINUTE 26

1. $\dfrac{1}{11} + \dfrac{6}{11} - \dfrac{2}{11} =$

2. When you divide fractions you should _______.
 a. invert the first fraction and then multiply
 b. invert the first fraction and then divide
 c. invert the second fraction and then multiply
 d. invert the second fraction and then divide

3. $13.467 \div 100 =$

4. $3.1 \cdot 4 =$

5. Complete the factor tree.

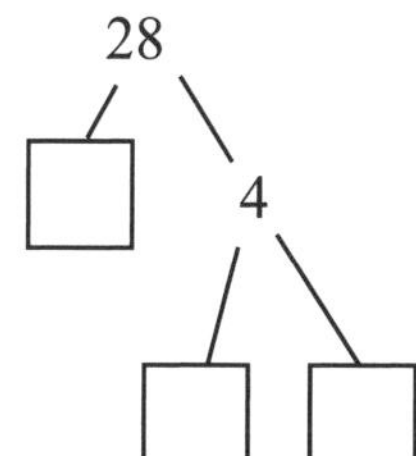

6. If you multiply the numbers in the three empty boxes in Problem 5 together, what do you get? _______

For Problems 7–9, use the chart to the right.

7. Who had the highest test score on the Chapter 5 test?

____

8. The difference between the highest and lowest scores (range) is about:
 a. 40 **b.** 25 **c.** 10 **d.** 15

9. Which one of the following is a reasonable average score (mean)?
 a. 95 **b.** 60 **c.** 70 **d.** 85

10. $6^2 - 8 = 28$ Circle: True or False

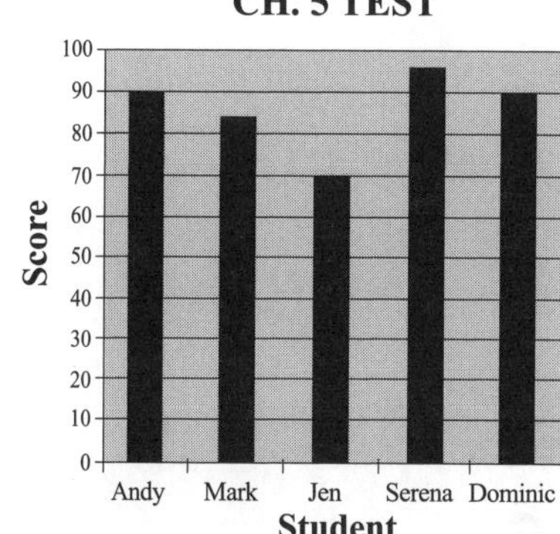

MINUTE 27

1. $\left[\dfrac{3}{5}\right]\left[\dfrac{2}{5}\right] =$

2. Reduce: $\dfrac{10}{40} =$

3. Circle the numerator and put a box around the denominator: $\dfrac{4}{15}$

4. There are two pictures on a wall. One is 12 in. × 4 in. and one is 9 in. × 6 in. Which one is larger? ___________________

5. To find the area of a shape, multiply the length by the width by the height.
Circle: True or False

6. How many quarters are in eight dollars? _______

7. Which of these could be the length of a bandage?
a. 3 inches   **b.** 3 meters   **c.** 3 millimeters   **d.** 3 kilometers

For Problems 8–10, use >, <, or = .

8. 10% of 200 _______ 50% of 100

9. 1^{99} _______ $0.\overline{9}$

10. $\sqrt{51}$ _______ 7

MINUTE 28

1. $4 \cdot 10 + 12 =$

2. $5 + 68 \div 4 =$

3. Find the perimeter of the parallelogram. ______

3.1 cm

2 cm

4. Complete the chart.

3^1	3
3^2	3·3
3^3	3·3·3
	3·3·3·3
3^5	

5. Circle all of the following that represent a form of multiplication.

$xy \qquad x(y) \qquad \dfrac{x}{y} \qquad (x)(y) \qquad x \div y \qquad x \cdot y$

6. The area and perimeter of the square to the right have the same numerical value. Circle: True or False

4

7. Circle the better deal: Ten donuts for \$2 or Two dozen donuts for \$6

8. To simplify $4 \cdot 3 - 3^2 + 1 \cdot 8$, which operation should be done first?

a. $4 \cdot 3$ **b.** $3^2 + 1$ **c.** 3^2 **d.** $1 \cdot 8$

9. Draw a horizontal line of symmetry.

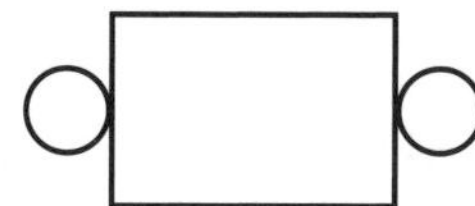

10. What is the pattern of these shapes? ______________________________

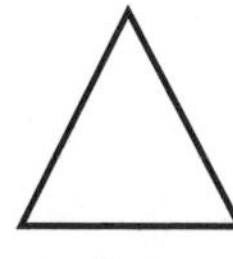
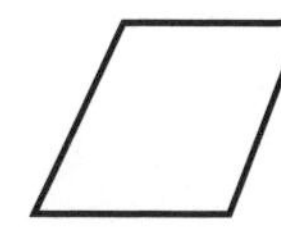
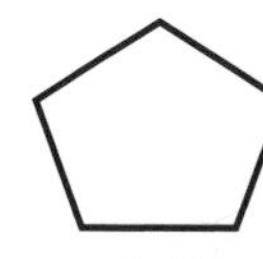

1st   2nd   3rd   4th

MINUTE 29

1. $0.35 + 0.4 + 0.1 =$

2. $0.2 \times 0.3 =$

3. Find the perimeter of the rectangle. _________

8.4 cm

5 cm

4. How many dots would the next shape in the sequence have? _______

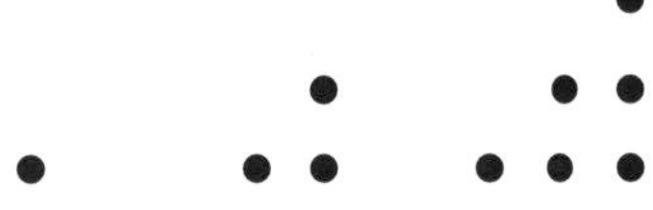

5. $192 + 206 \approx$ _______. (**Hint:** "$\approx$" means "approximately")
  a. 500 **b.** 300 **c.** 200 **d.** 400

For Problems 6–10, match the words with their correct algebraic expression.

6. nine divided by n plus two **a.** $4n$

7. n plus nine squared **b.** $\dfrac{4}{n-9}$

8. four times the sum of nine plus n **c.** $\dfrac{9}{n}+2$

9. the product of four and n **d.** $4(9+n)$

10. four divided by the difference of n and nine **e.** $n+9^2$

MINUTE 30

1. Laurie says that $2 + 3 \times 2 + 3 = 13$. Ray says that $2 + 3 \times 2 + 3 = 11$. Who is correct? ____________________

2. The first step in simplifying $400 - 5(12 + 13)$ would be to _______.
   a. add   **b.** subtract   **c.** multiply   **d.** divide

3. Insert parenthesis ( ) to make the following problem true: $3 + 6 - 2 \bullet 4 = 19$

4. Does $a = 4$ solve the equation $5a - 3 = 17$?   Circle: Yes   or   No

5. In the grid to the right, circle a diagonal sum that equals 15.
   (**Hint:** Look for three numbers.)

4	2	7	8
9	6	4	4
3	5	5	1
2	8	3	8

6. Circle all the numbers that make the inequality $a + 2 < 7$ true.
   2   3   4   5   6   7

7. If $x + \dfrac{2}{2} = \dfrac{5}{2}$, then $x =$ _______.

For Problems 8–10, shade the box with the correct equivalent.

8. 1 mile =

5,280 feet	454 grams	2.54 inches	1 kilometer

9. 1 ton =

16 ounces	2,000 pounds	454 grams	1,000 milligrams

10. 1 gallon =

2 cups	1 liter	1,000 milliliters	4 quarts

MINUTE 31

1. Fill in the missing numbers.

$$\begin{array}{r} 9.36 \\ +\,1.0\,\square \\ \hline \square\,0.41 \end{array}$$

2. $21 \cdot \dfrac{1}{3} =$

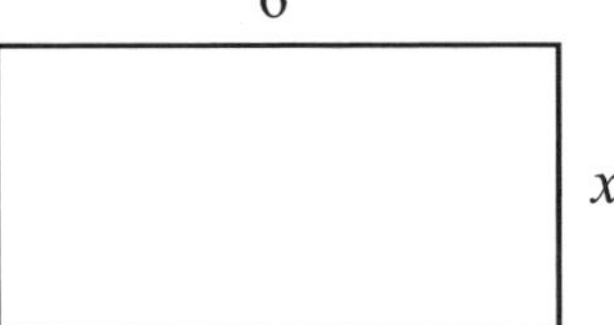

3. Find x if the perimeter of this rectangle is 20. _______

For Problems 4–6, use the grid to the right.

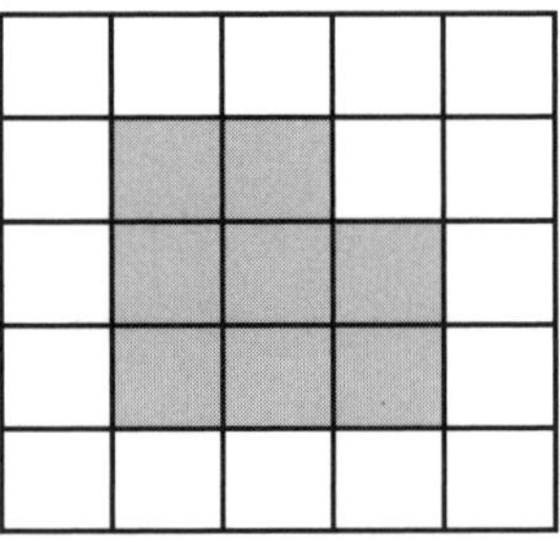

4. What is the area of the shaded region? _______

5. What is the perimeter of the shaded region? _______

6. What percentage of the boxes are shaded? _______

7. Circle the numbers that make $\dfrac{n}{5} \le 3$ a true statement:

5   10   15   20

8. If the time is 4:15, what time will it be in nine hours? _______

9. If you rearranged the numbers in 1,996, what is the largest number you can make? _______

10. Shade the shape with the most right angles.

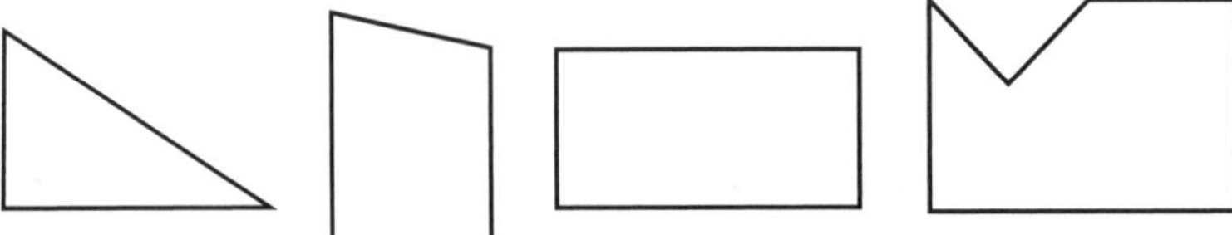

MINUTE 32

For Problems 1–2, use the box to the right.

9	2	
1		8
	7	3

1. Using the numbers 4, 5, and 6, fill in the empty boxes so the rows and columns add up to 15.

2. Do the diagonals in Problem 1 also add up to 15?
Circle: Yes   or   No

For Problems 3–5, use the calendar to the right.

MARCH

S	M	T	W	T	F	S
				1	2	3
4	5	6	7	8	9	10
11	12	13	14	15	16	17
18	19	20	21	22	23	24
25	26	27	28	29	30	31

3. What date is two weeks from the 5th? ______________

4. If apartment rentals cost \$10 per day, how much will it cost to rent an apartment for the month of March?

____

5. How many weekend days are there in March? ________

6. Roger has successfully caught 10 passes in a row. What conclusion can we make about his next (11th) attempt?
a. Roger will catch the 11th pass.
b. Roger will drop the 11th pass.
c. Roger may catch or drop the 11th pass.

7. If $\dfrac{12}{20} = \dfrac{x}{100}$, then $x =$ _______.

For Problems 8–10, cross out the item that does NOT belong in each list.

8. gallons   liters   cups   grams

9. miles   feet   inches   meters

10. pounds   centimeters   grams   ounces

MINUTE 33

1. Complete the times table to the right.

×	7	8
5		40
6	42	

2. Seven quarters, three dimes, and one nickel = $ _______ .

3. If $a + 12 = 31$, then $a =$ _______ .

4. The sum of two identical numbers is 16. What is the number? _______________

For Problems 5–6, use the number line to the right.

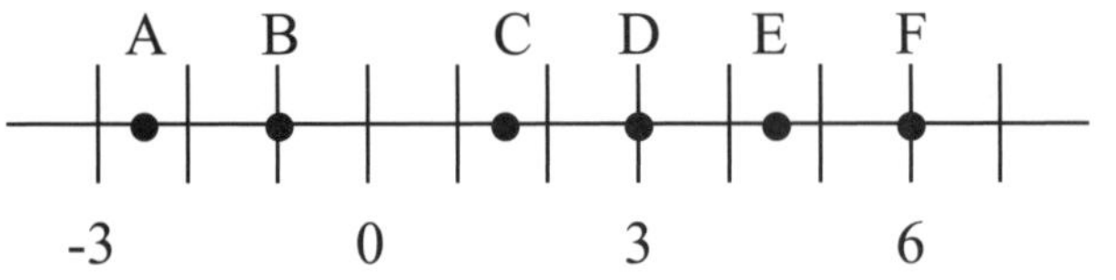

5. Which letters represent fractions?

____

6. Which letter is located directly between 3 and 6? _______

For Problems 7–10, cross out the item that does NOT belong in each list.

7. 2   6   10   11

8. 3   7   12   13

9.

10. 65% $\dfrac{2}{3}$ 0.$\overline{6}$

MINUTE 34

1. $20(300) =$

2. $2\overline{)36} =$

3. Complete the missing numbers in the table to the right.

Sum	Product	Numbers
12	35	__ and __

4. Which line segment is longer?
Circle: $\overline{XY}$ or $\overline{YZ}$

5. Using the line given in Problem 4, find $\overline{XZ}$ if $\overline{XY} = 6$ and $\overline{YZ} = 3$. _______

6. Shade 75% of this circle.

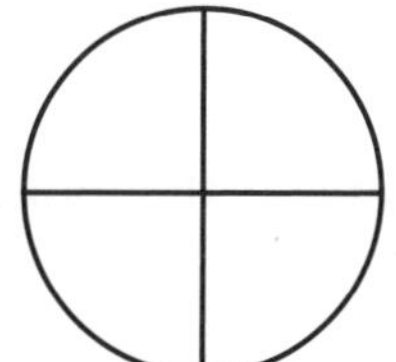

7. Write as an improper fraction: $5\frac{1}{3} =$

8. What number does point A represent?

____

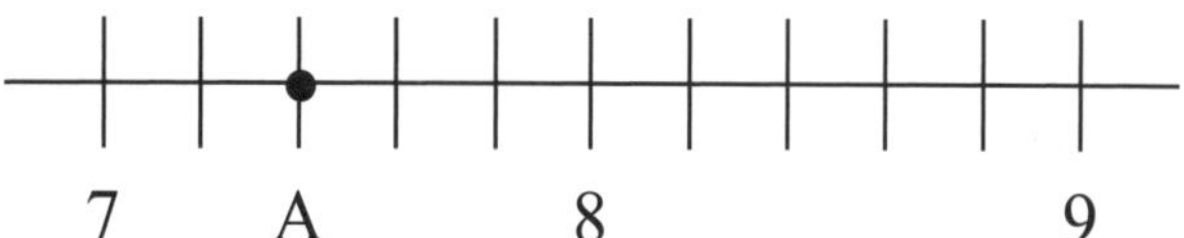

9. $\frac{3}{5} \cdot \frac{1}{2} \cdot \frac{3}{4} =$

10. A tile on the floor looks similar to the shape to the right.
If an egg is accidentally dropped on the tile, where would
it be more likely to land?
Circle: A gray square   or   A white square

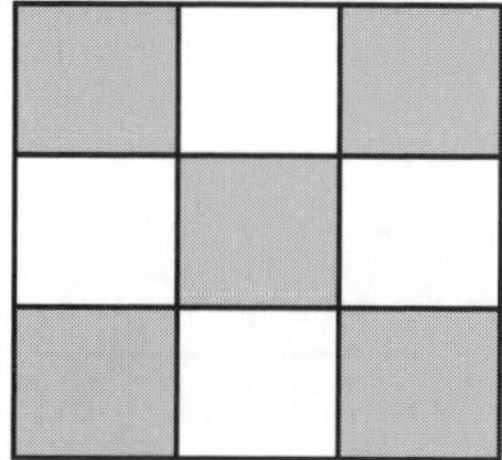

MINUTE 35

1. $\begin{aligned} \$\,40.75 \\ -\ 4.57 \end{aligned}$

2. If $15 \times a = 135$, then $a =$ _______ .

For Problems 3–4, use the grid at the right.

3. What fraction of the rectangle is shaded?
(express in lowest terms) _______

4. What fraction of the rectangle is NOT shaded?
(express in lowest terms) _______

5. Which one of the following line segments is the longest?
a. $\overline{AB}$ **b.** $\overline{BC}$ **c.** $\overline{AC}$

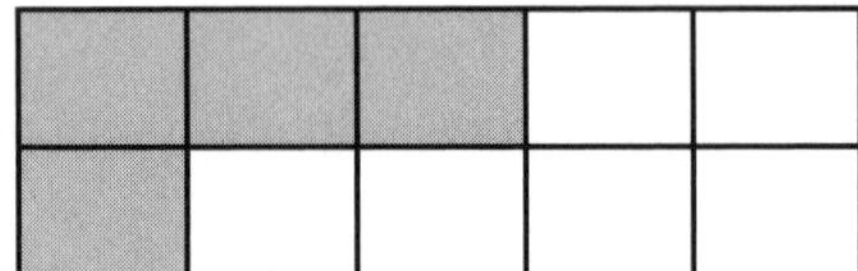

6. Using the number line given in Problem 5, if $\overline{AC} = 12m$ and $\overline{BC} = 7m$,
then $\overline{AB} =$ _______ .

For Problems 7–10, cross out the item that does NOT belong in each list.

7. 5   7   11   14

8. 5   9   27   63

9. $\dfrac{5}{5}$ 1^9 1% $\sqrt{1}$

10. B L A C K   B R O W N   G R E E N   R E D

MINUTE 36

1. Write $\dfrac{7}{4}$ as a mixed number. _______________

2. Reduce: $\dfrac{4}{20} =$

3. What is the reciprocal of $\dfrac{8}{3}$? _______

4. If $a = 28$ and $b = 4$, then $\dfrac{a}{b} =$ _______.

5. Ten dollars is equal to _______ pennies.

For Problems 6–7, use the triangle to the right.

6. What is the perimeter of the triangle? _______

7. What is the area of the triangle? _______________
(**Hint:** Take half of the base times the height.)

8. Use the information below to fill in the Venn diagram to the right.

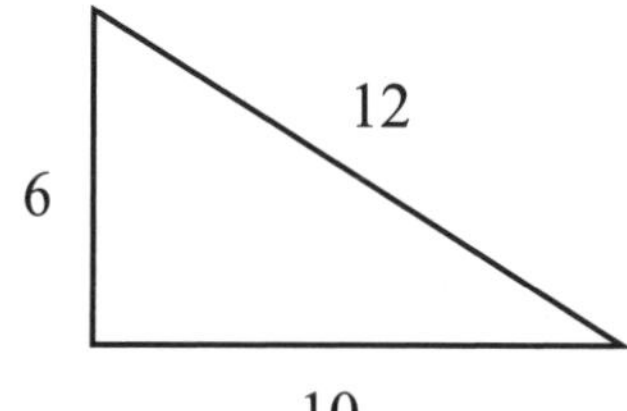

Students' Favorite Foods

Hamburgers	8
Hot dogs	10
Both	4

9. These two lines are _______________.
Circle: parallel   or   perpendicular

10. Find the next letter and number in the series: A4, D6, G8, J10, _______

MINUTE 37

1. If $\dfrac{1}{3} = \dfrac{x}{6}$, then $x =$ _______.

2. $\dfrac{1}{3} + \dfrac{3}{6} =$

3. $0.46 + 0.05 =$

4. Fill in the missing number: $5 \bullet \boxed{} = 0.25$

5. Which of these numbers represents seventeen thousandths?
   a. 0.0017   **b.** 0.17   **c.** 0.017   **d.** 0.00017

6. Put the numbers {4, 12, 16, 18, 20} into the Venn diagram.
   (**Hint:** One of the numbers will go in both rings.)

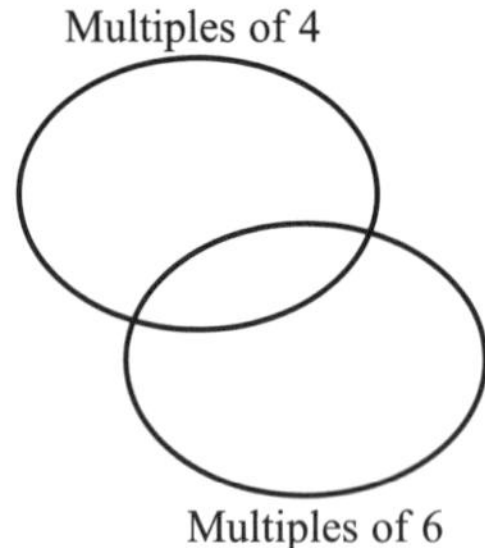

7. Shade the boxes in the 4th shape to create the next shape in the sequence.

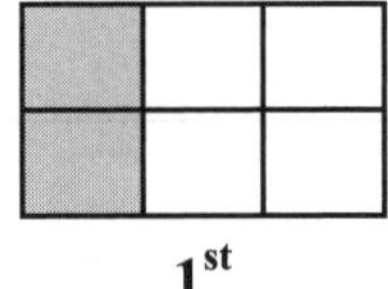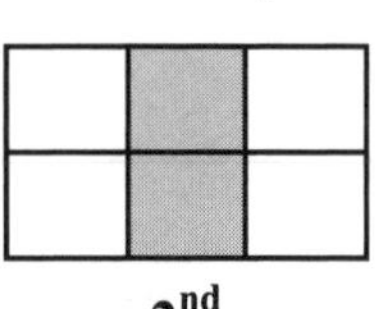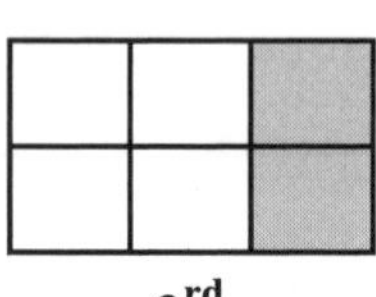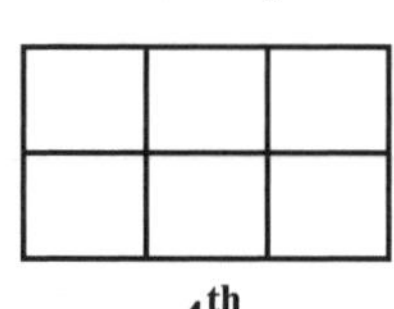

8. Find the two prime numbers that complete the equation. $\boxed{} + \boxed{} = 12$

9. Draw the horizontal and vertical lines of symmetry in this figure:

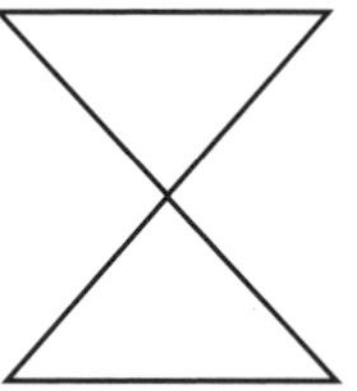

10. $2\dfrac{1}{4}$ km = _______ meters

MINUTE 38

1. $\dfrac{1}{5} - \dfrac{1}{10} =$

2. Circle three consecutive decimals in the grid that have a sum of 0.8. (**Hint:** no diagonals)

0.3	0.4	0.9
0.2	0.2	0.2
0.6	0.2	0.5

3. $0.3(5 + 3 - 2) =$

4. Circle the net below that will create a triangular pyramid.

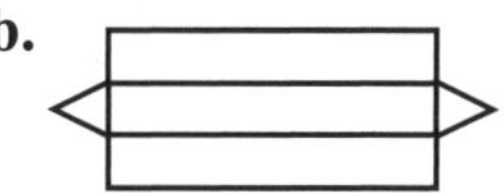 a.   b.   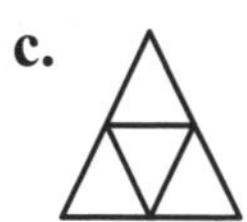 c.   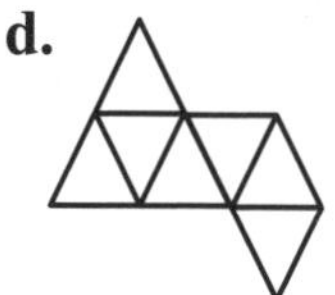 d.

5. Write the number twenty-three thousandths. _______________

6. Round the number $45.\overline{6}$ to the nearest tenth. _______

For Problems 7–10, cross out the item that does NOT belong in each list.

7. $\dfrac{1}{2}$ 0.5   50%   0.05

8. length × width $\dfrac{1}{2}$ length × width   base × height   length × width × height

9. 81   20   36   49

10. red   yellow   orange   purple

MINUTE 39

1. In the number 38.7165, what number is in the hundredths position? _______

2. Round the number in Problem 1 to the nearest thousandth. _____________

3. The least common denominator of $\dfrac{1}{4}$ and $\dfrac{1}{6}$ is _______.

For Problems 4–5, use the picture to the right.

4. If the black dots represent Beth's three "hits," what is her score on the dartboard? _______

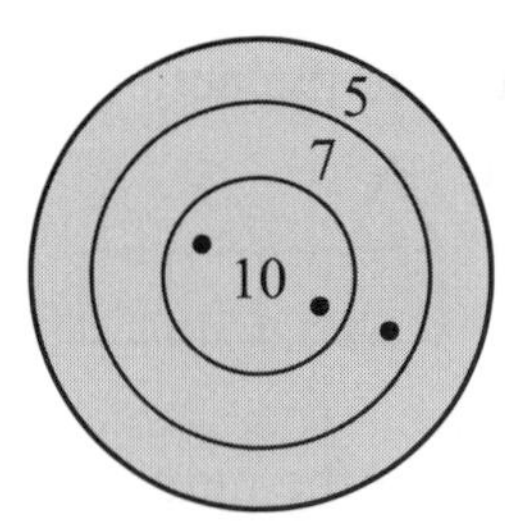

5. If Beth "hits" a 5 on her next throw, what will her total be? _______

6. Find the next letter and number in the series: Z1, Y2, X3, W4, _______

For Problems 7–10, match the words with their correct algebraic expression.

7. nine times n plus 1   **a.** $9(n + 1)$

8. the square root of n **b.** $\dfrac{n}{9}$

9. nine times the sum of n and 1   **c.** $9n + 1$

10. the quotient of n and 9   **d.** $\sqrt{n}$

MINUTE 40

1. Complete the times table.

$\times$	9	10
12		120
13	117	

2. Order the decimals {0.058, 0.508, 0.085, 0.580} from least to greatest.

____

3. Draw a dot at the midpoint of A and B and label it C.

4. Using the line in Problem 3, if $\overline{AB}$ = 11, then $\overline{AC}$ = _______.

For Problems 5–7, use the frequency table to the right.

5. What was the highest score? _______

6. What score occurred most often? _______

7. How many people took the test? _______

Score	Tally
95	I I
90	I I I I
85	
80	I I I
75	I
70	I I
65	I
Below 60	I I

8. The length and width of a box are 4 in. The volume is 48 in.^{3}
What is the height of the box? _______

9. Circle the numbers that are greater than 1,100.

1,109 10^4 1,006   999 $\sqrt{1\ billion}$

10. $4 \cdot 6 \cdot 8 \cdot 0 \cdot 5 \cdot 2 =$

MINUTE 41

1. Order the decimals {3.0, 0.3, 0.33, 3.3} in ascending order (least to greatest).

____

2. Fill in the remaining factors of 30.

1		3	5		10		30

For Problems 3–5, use the chart to the right.

Exercise Day	Tally (hundreds)
M	I I I I
T	I I
W	I I I
TH	I
F	I I
S	I I I I I I
SU	I I

3. More people exercised on _________________ than any other day.

4. Fewer people exercised on _________________ than any other day.

5. On Saturday, _______ times as many people exercised than on Friday.

For Problems 6–7, use the Venn diagram to the right.

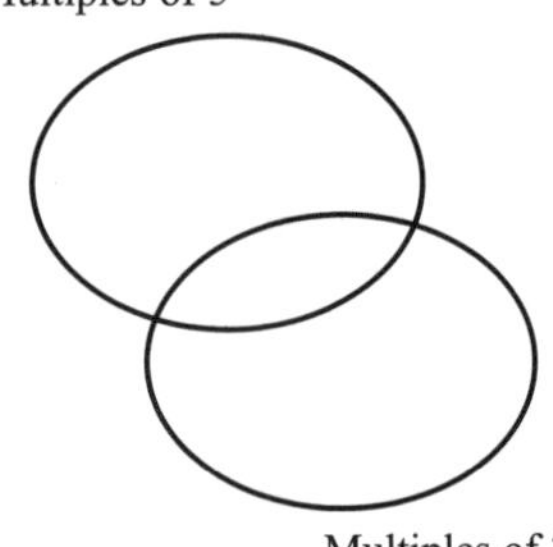

6. Put the numbers 5, 14, 20, 21, 30, and 35 into the Venn diagram.

7. Which number from Problem 6 belongs in both circles? _______

For Problems 8–10, evaluate the expressions if $a = 4$, $b = 6$, and $c = 10$.

8. $\dfrac{5b}{c} =$

9. $\dfrac{1}{2}ab =$

10. $a(b + c) =$

MINUTE 42

1. Can 233 be evenly divided by 2?   Circle: Yes   or   No

2. What is the rule for the following sequence: 16, 24, 36, 54, 81, . . .?
a. add 12   **b.** add 18   **c.** multiply by 1.5   **d.** multiply by 2

3. Complete the table.

Fraction	Decimal	Percent
	0.3	30%

For Problems 4–7, use the circle graph to the right.

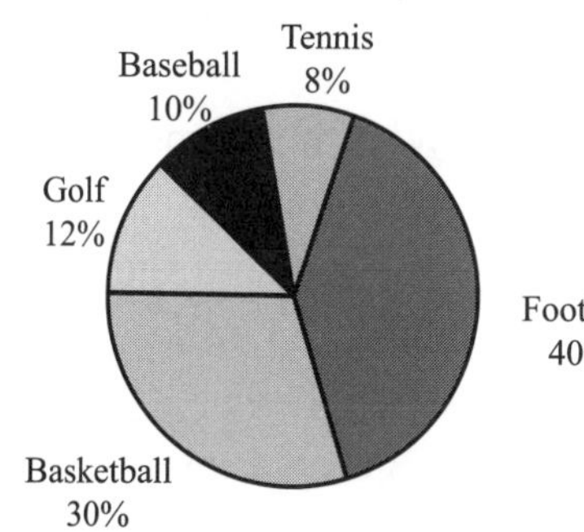

4. Which is the more popular sport: golf or tennis?

____

5. What two sports added together have the same percentage as football?

________________   ________________

6. Which two sports added together represent half of everyone surveyed? ________________   ________________

7. If 300 people took part in this survey, then _______ people said that baseball was their favorite sport.

8. Is the number $\frac{1}{6}$ closer to 0, $\frac{1}{2}$, or 1? _______

9. Is the dashed line shown a line of symmetry? 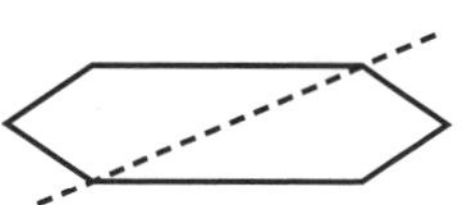
Circle: Yes   or   No

10. Does $n = 7$ solve the problem $2n + 3.5 = 17.5$?   Circle: Yes   or   No

MINUTE 43

1. Shade 15% of the boxes.
(**Hint:** 5% are already shaded for you)

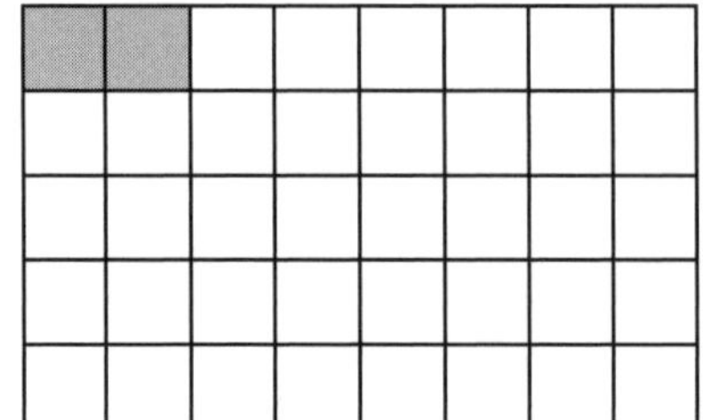

2. $16.29 - 0.3 =$

3. $2 + 0.2 + 0.02 + 0.002 =$

4. There are 20 nickels in a dollar. How many nickels are in 25 dollars? _______

For Problems 5–8, use the frequency table to the right.

Score	Tally
95	I
90	I I
85	I I I I I
80	
75	I
70	I I
65	I
Below 60	I

5. What is the mode? _______

6. The mean of the scores is 80. If Sarah gets a 90, the mean will _______.
a. go down   **b.** stay the same
c. go up a lot   **d.** go up a little

7. The median (score in the middle) is _______.

8. How many people took the test? _______

9. Which of the following is the next shape in the pattern?

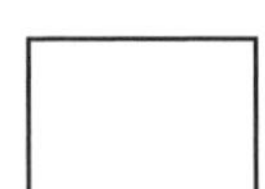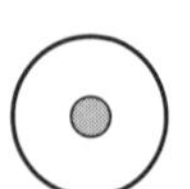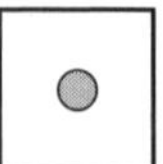

a.   **b.**

c.   **d.**

10. Put a decimal point in the number 26583 so that the 5 has a value of $\dfrac{5}{100}$. _______

MINUTE 44

1. $\dfrac{7}{12} - \dfrac{1}{2} =$

For Problems 2–3, use the grid to the right.

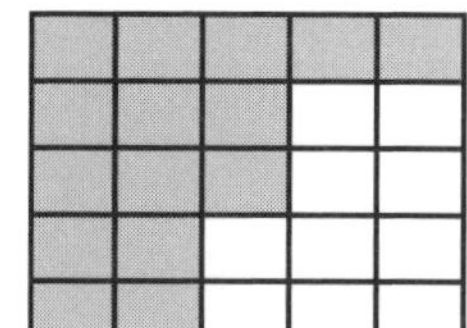

2. What fraction of the squares are shaded? (Write in lowest terms.) _______

3. What fraction of the squares are NOT shaded? _______

For Problems 4–5, use the chart to the right.

M	T	W	T	F	Sat	Sun
32	16	8				

4. Mary started the week with 32 bananas. On Tuesday her family ate half of them. On Wednesday they ate half of the remaining bananas. If they continue doing this each day, on which day of the week will only one banana be left? ___________

5. If Mary's family continues to eat half of the remaining banana supply each day, will they ever get to zero bananas?   Circle: Yes   or   No

For Problems 6–8, fill in the boxes to complete the equivalencies.

6. 1 m = 100 cm   3 m = ☐ cm   60 cm = ☐ m

7. 1 kg = 1,000 g   3.2 kg = ☐ g   60 g = ☐ kg

8. 1 yd. = 36 in.   3 yd. = ☐ in.   144 in. = ☐ yd.

9. What number does Point A represent? _______

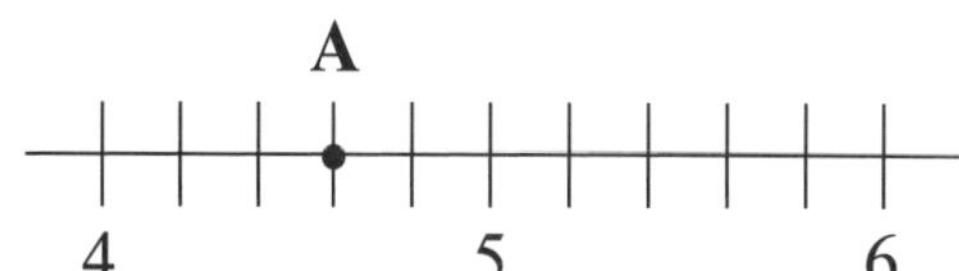

10. Cross out the shape that does NOT belong.

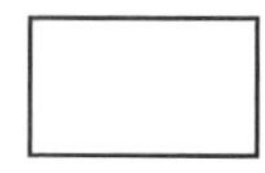   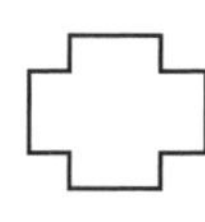   

a.   b.   c.   d.   e.

MINUTE 45

1.
$$10.38 + 1.26$$

2.
$$3.4 \times 0.2$$

3. $0.2 + 0.3 + 0.5 + 0.2 =$

4. These lines are _______.
Circle: parallel   or   perpendicular

5. What number is the arrow pointing toward in the number line to the right? _______

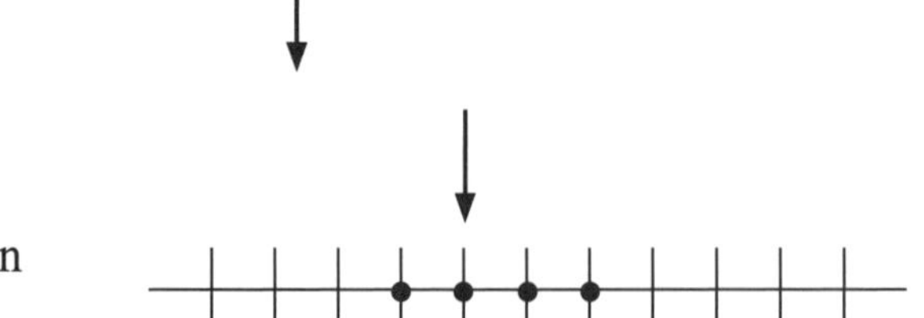

6. Circle the number that is different from the others.

226   357   486   451   842

For Problems 7–10, circle *True* or *False* if $a = 3$, $b = 5$, and $c = 11$.

7. a, b, and c are prime numbers   True   or   False

8. $ab > bc$ True   or   False

9. $a^b = b^a$ True   or   False

10. $a + b + c =$ a prime number   True   or   False

MINUTE 46

1. $9\overline{)729} =$

2. Put a decimal in the number 3467 so that the 7 has a value of $\dfrac{7}{100}$. ____________

3. Fill in the remaining composite numbers between 4 and 18.

4	6				12		15	16	18

4. A regular polygon is a shape with all sides equal in length. Which of these is an irregular polygon?

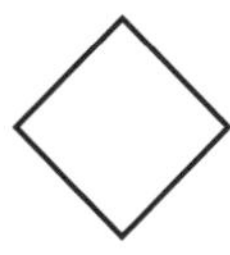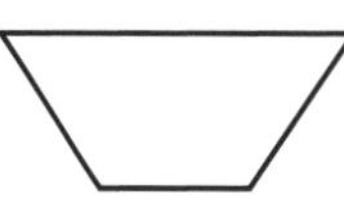

   a.   **b.**   **c.**   **d.**

5. Bill has $3. Tom has twice as much as Bill. Linda has three times as much as Tom. How much does Linda have? _______________

6. Draw perpendicular diameters in the circle. 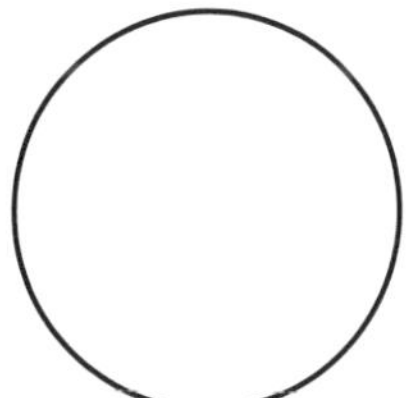

7. If $\dfrac{2}{5} = \dfrac{a}{10}$, then $a =$ _______.

8. Use the digits 4, 9, and 1 to make two numbers greater than 875. _______ _______

9. What numbers in the set {2, 4, 6, 8, 10} satisfy the inequality $\dfrac{n}{2} + 1^3 \geq 5$? ____________

10. Shade the 2nd circle after the 3rd circle from the left.

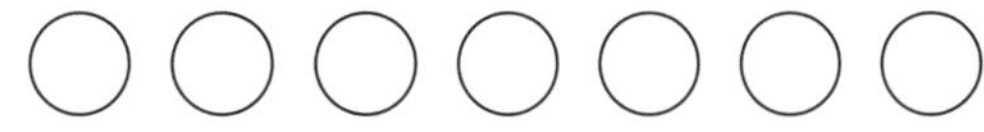

MINUTE 47

1. 132 minutes = _______ hour(s) _______ minutes.

For Problems 2–4, use the circle graph to the right.

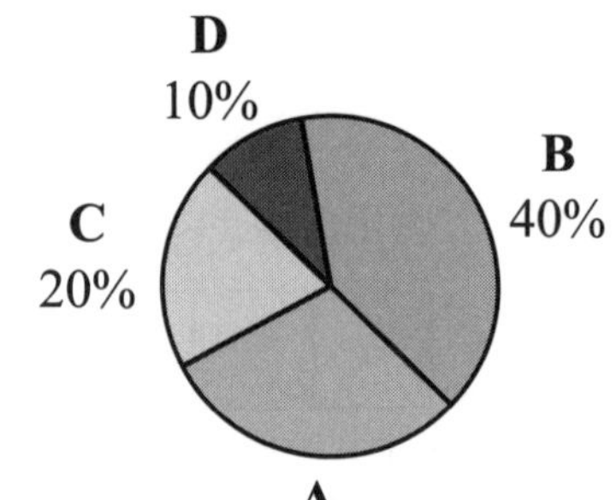

2. What percent must category A be equal to? _______

3. Which two categories make up 50% of the graph?
_______ and _______.

4. If these were the grades on a recent test, then the majority
of the class_______.   Circle: Passed   or   Failed

5. $\left[\dfrac{1}{3}\right]\left[\dfrac{1}{4}\right] + \left[\dfrac{2}{3}\right]\left[\dfrac{3}{4}\right] =$

For Problems 6–10, match each word with its correct definition.

6. perpendicular   **a.** A number that can only be divided by 1 and itself.

7. parallel   **b.** Two lines that never intersect and are spaced equally apart.

8. diameter   **c.** Two lines that intersect at right angles.

9. prime   **d.** The distance across a circle through its center.

10. composite   **e.** A number having other factors besides 1 and itself.

MINUTE 48

For Problems 1–3, use the figure to the right.

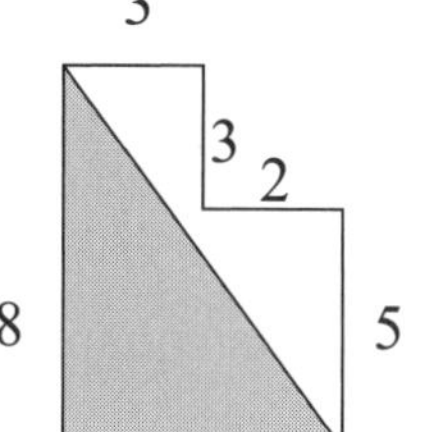

1. What is the width of the base of the hexagon? _______

2. What is the perimeter of the hexagon? _______

3. What is the area of the shaded triangle? _______

4. 10% of 120 =

5. If $8m = 416$, then $m =$ _______.

For Problems 6–10, match each word with its correct definition.

6. factor   **a.** a six-sided shape

7. hexagon   **b.** the amount of surface a shape covers

8. pentagon   **c.** the distance around the outside of a shape

9. perimeter   **d.** a number that goes evenly into another number

10. area   **e.** a five-sided shape

MINUTE 49

1. $8\overline{)32.16}$ =

2. Fill in the square to complete the equation. $\square \cdot \dfrac{1}{4} = \dfrac{3}{16}$

3. 15 seconds = _______ minutes.   Circle: 4   0.5   2   0.25

4. What is the perimeter of this rectangle? _______

2

4.4

5. What is the area of the rectangle in Problem 4? _______

6. Do all rows and columns add up to the same number in this grid?
Circle: Yes   or   No

3	8	4
9	1	5
2	7	6

7. Fill in the missing number in the box.

2
5 ⟶ 8 ⟶ 11
8 ⟶ 32 ⟶ ☐

For Problems 8–10, estimate to find the best answer.

8. 26 out of 99 =
a. 10%   **b.** 40%   **c.** 75%   **d.** 25%

9. 11% of 80 =
a. 8   **b.** 0.8   **c.** 20   **d.** 79

10. $\dfrac{29}{50}$ =
a. 29%   **b.** 60%   **c.** 14%   **d.** 200%

MINUTE 50

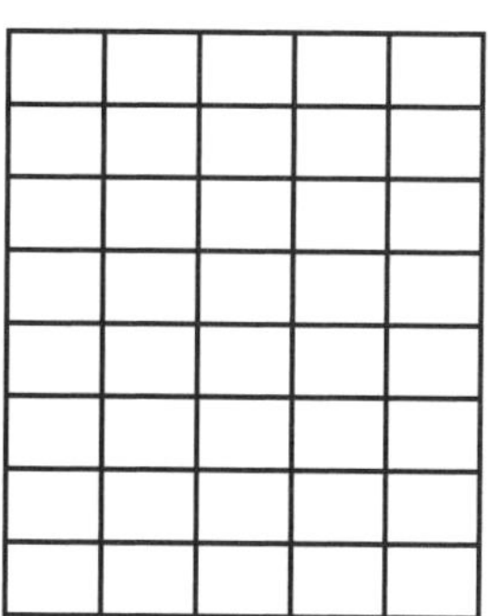

For Problems 1–3, use the grid to the right.

1. Shade 15% of the squares.

2. What percent of the squares will NOT be shaded? _______

3. What is the perimeter of the grid? _______

4. Shade the squares in the 4th shape to complete the sequence.

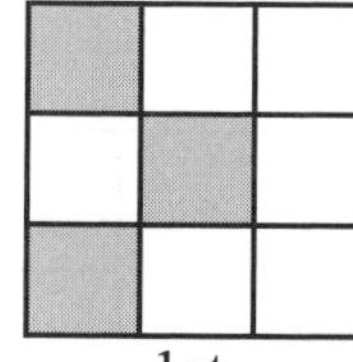 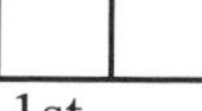 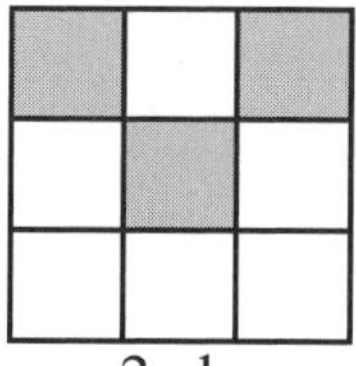 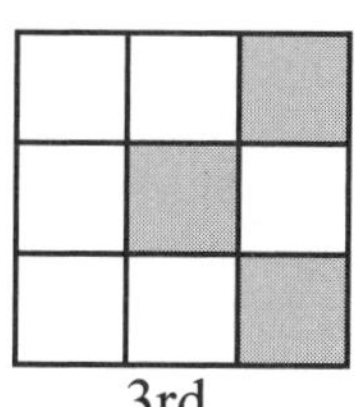 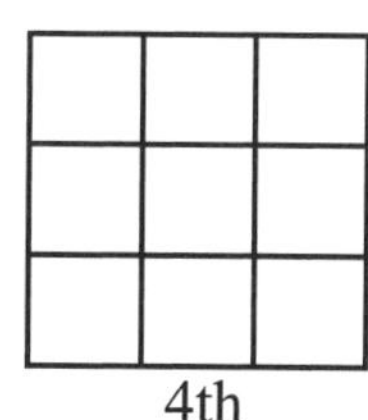

1st   2nd   3rd   4th

5. The ages of the Eagle Cadet group members are 4, 6, 7, 7, and 11. What is the mode age? _______

6. What is the mean age of the Cadet group in Problem 5? _______

7. What is the median age of the Cadet group in Problem 5? _______

8. $3 + 6^2 \div 12 =$

9. If $y = 3x - 6$ and $x = 7$, then $y =$ _______.

10. $2^2(3 + 7 - 1) =$

MINUTE 51

Rules of Integers
$$(-)(-) = +$$
$$(-)(+) = -$$
$$(-) \div (-) = +$$
$$(-) \div (+) = -$$
$$(-) + (-) = -$$

1. $-7 \cdot -8 =$

2. $-6 \cdot 7 =$

3. According to the chart, a negative plus a negative makes a _______________.

4. $(-5)^2 =$

5. If $\dfrac{12}{n} = 24$, then $n =$ _______.

6. Use the function rule above the chart to fill in the empty boxes.

$y = 2x - 3$

x	y
4	
5	7
10	

7. $3.426 \times 10^3 =$

8. What is the volume of the box? _____________

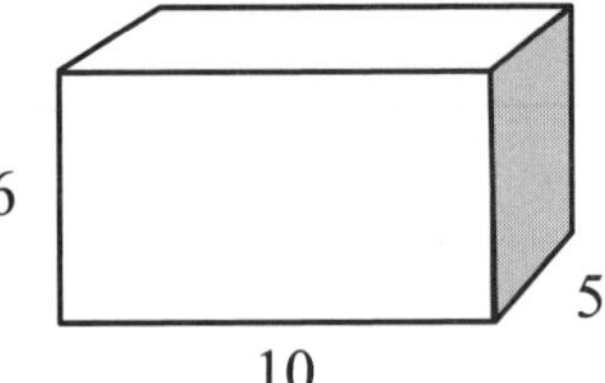

9. A bag holds seven red marbles and three blue marbles. If Jill reaches into the bag and pulls out one marble, what is the probability that the marble will be red? _____________

10. If all 10 marbles described in Problem 9 were still in the bag, what is the probability that Jill would pull out a blue marble? _____________

MINUTE 52

1. $\dfrac{-45}{9} =$

2. $(-5) + (-8) =$

3. $(-2 \cdot -4)^2 =$

Rules of Integers
$(-)(-) = +$
$(-)(+) = -$
$(-) \div (-) = +$
$(-) \div (+) = -$
$(-) + (-) = -$

x	y
2	13
5	28
3	18

4. Look at the chart and complete the function rule.
$y = 5x +$ _________

5. How many small blocks make up this shape? _________
(**Hint:** be sure to count the blocks you can't see)

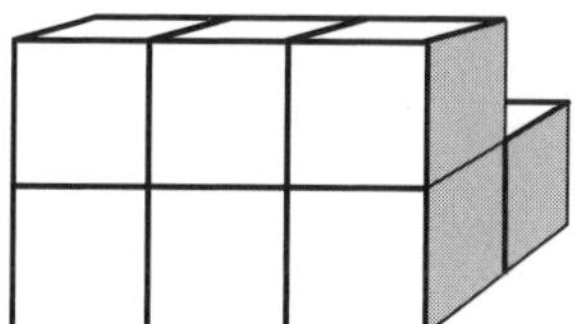

6. What number on the number line is the arrow pointing toward? _________

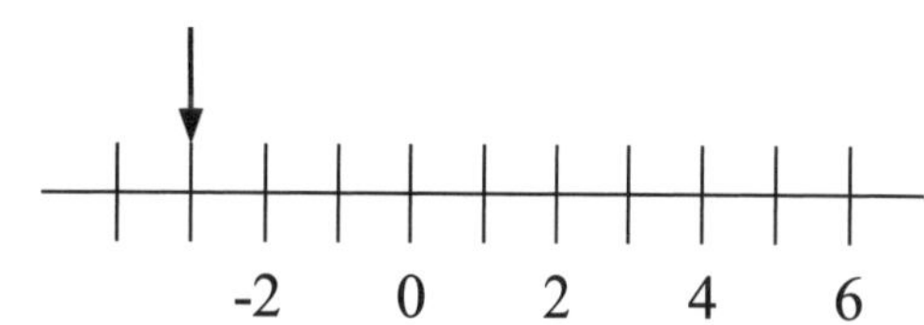

For Problems 7–10, evaluate if $x = -2$, $y = 3$, and $z = 10$.

7. $xyz =$

8. $2xy =$

9. $\dfrac{y}{z} =$ _________ %

10. $\dfrac{z}{y + 2} =$

MINUTE 53

1. If $8n = -40$, then $n =$ _______.

2. If $\dfrac{n}{4} = 12$, then $n =$ _______.

For Problems 3–5, use the chart to the right.

3. $y_2 - y_1 =$

y_1	y_2	x_1	x_2
6	12	3	5

4. $x_2 - x_1 =$

5. $\dfrac{y_2 - y_1}{x_2 - x_1} =$

For Problems 6–10, use the coordinate grid to the right.

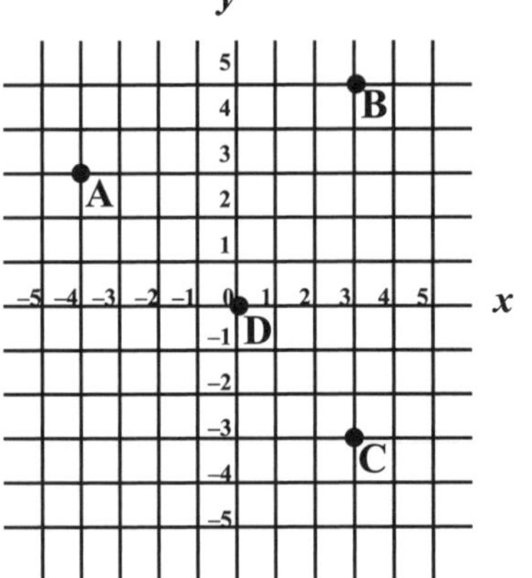

6. Which letter is at the origin (0, 0) of the grid? _______

7. Which letter(s) are located three units to the right of the origin? _____________

8. Which letters are located above the origin? _____________

9. To go from point A to point B you would have to go _______.
a. NE   **b.** SE   **c.** SW   **d.** NW

10. Is there a letter located four units left of the origin and down two units?
Circle: Yes   or   No

MINUTE 54

1. $3 + (-4)(-3) - 5 =$

Rules of Integers

$(-)(-) = +$
$(-)(+) = -$
$(-) \div (-) = +$
$(-) \div (+) = -$
$(-) + (-) = -$

2. $\dfrac{(-5) + (-13)}{6} =$

3. If $-7m = -28$, then $m = $ _______.

4. Look at the chart and complete the function rule.

$y = x^2 +$ _______

x	y
1	2
2	5
5	26

5. Using the chart in Problem 4, if $x = 10$, then $y = $ _______.

For Problems 6–8, use the chart to the right.

6. $y_2 - y_1 =$

y_1	y_2	x_1	x_2
4	10	2	5

7. $x_2 - x_1 =$

8. $\dfrac{y_2 - y_1}{x_2 - x_1} =$

9. Put the numbers {10, -10, 5, -5, 0} in ascending (smallest to greatest) order.

____

10. Put the numbers {-5, 0, 3^2, $(-2)^2$} in descending (greatest to smallest) order.

____

MINUTE 55

1. How many blocks are in the shape to the right? _______ 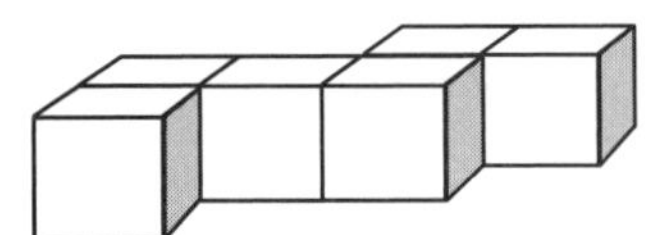

2. Shade the squares in the 4th shape to complete the sequence.

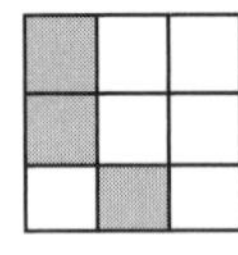 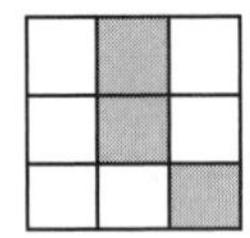 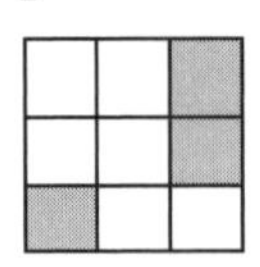 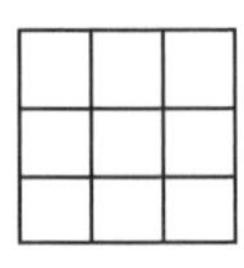

    1st   2nd   3rd   4th

3. Shade the octagon.

 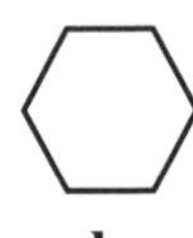 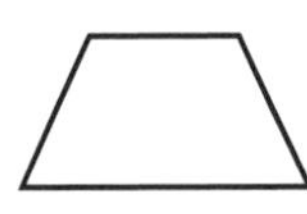 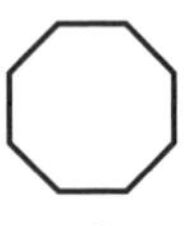

   a.   **b.**   **c.**   **d.**

4. Shade the trapezoid.

  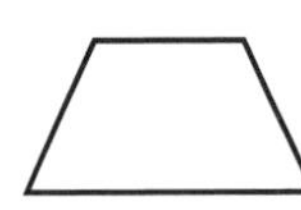 

   a.   **b.**   **c.**   **d.**

For Problems 5–8, use the coordinate grid to the right.

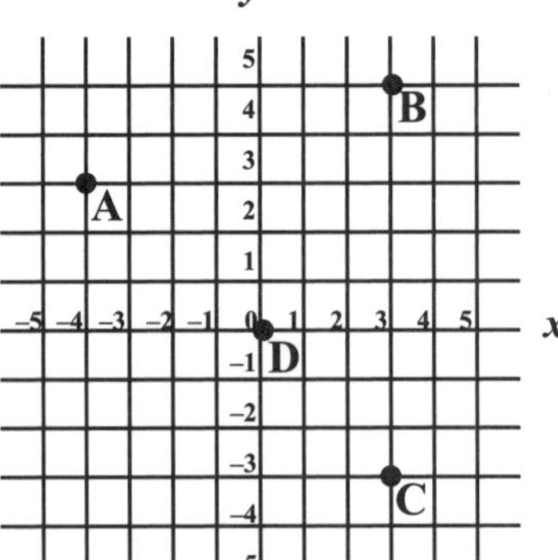

5. Which letter is at the origin (0, 0) of the grid? _______

6. The coordinates of point B are (3, 5). What are the coordinates of point C? _______

7. What are the coordinates of point A? _______

8. To go from point C to point A, you have to go _______.
  a. NE   **b.** SE   **c.** SW   **d.** NW

For Problems 9–10, use > , <, or = to complete.

9. $(-8)(-5)$ _______ $(9)(-8)$

10. $\dfrac{(-6)^2}{4}$ _______ $\sqrt{(-4)(-25)}$

MINUTE 56

1. Use •, +, −, or ÷ to complete: 5 ☐ 12 ☐ 3 = 9

2. If $\left[\dfrac{3}{13}\right]\left[\dfrac{a}{4}\right] = \dfrac{15}{52}$, then $a =$ _______.

3. If $36 = 2^x \cdot 3^x$, then $x =$ _______.

4. Write .01212… using bar notation. _______

5. If you multiply the number _____ times itself and add 1, you get 37.

6. Write $10\dfrac{3}{4}$ as an improper fraction. _______

For Problems 7–10, circle *True* or *False*.

7. Railroad tracks are a good example of perpendicular lines.   True   or   False

8. (negative) × (negative) × (negative) = positive.   True   or   False

9. The fraction $\dfrac{2}{3}$ is closer to $\dfrac{1}{2}$ than it is to 1.   True   or   False

10. Trapezoids, squares, and rectangles all have four sides.   True   or   False

MINUTE 57

1. $2(-5 + 3 \cdot 4) =$

2. If $3n - 2 = 10$, then $n =$ _______.

3. If $40 = 2^x \cdot 5$, then $x =$ _______.

For Problems 4–6, use the coordinate grid to the right.

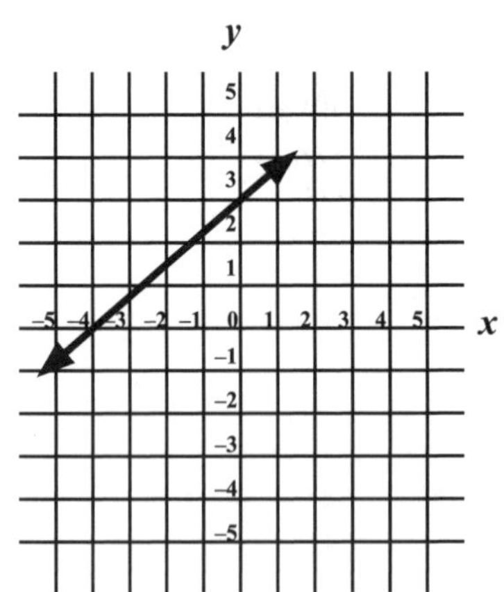

4. As you move from left to right, the line on the grid:
Circle: goes up   goes down   is level

5. Where does the line cross the y-axis? _______

6. Where does the line cross the x-axis? _______

7. Find the next letter and number in the series: A3, D6, G9, _______.

8. Look at the chart and complete the function rule.
$y =$ ___ $x + 2$

x	y
1	4
2	6
3	8

9. Using the chart in Problem 8, if $x = 10$, then $y =$ _______.

10. Ali flips a coin two times. The possible results are shown to the right. List the four possible outcomes for two flips. Two have been done for you.

HH, HT, _______, _______.

MINUTE 58

1. Use + or – to complete. (3 ☐ 6) ☐ 12 = 9

2. $(-3)^3 =$

3. If all the angles of a triangle total 180°, then angle x in this triangle is _______.

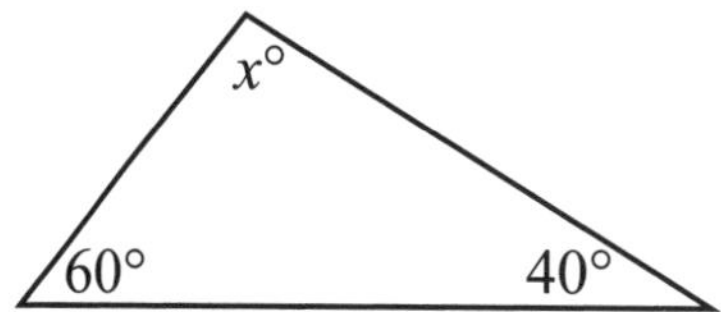

4. Martin folds a sheet of paper in half, then in half again, and in half yet again. When he unfolds it, the paper is divided into _______ sections.

5. This letter **H** has _______.
   a. parallel lines   **b.** perpendicular lines   **c.** both

6. A is to **A**, as ☐ is to _______.

  a. ☐   **b.** ☐   **c.** ☐   **d.** ☐

7. If point A, one of the vertices of a pentagon, is connected to each other vertex in the pentagon, _______ triangles will be formed.
  a. 2   **b.** 3
  c. 4   **d.** 5

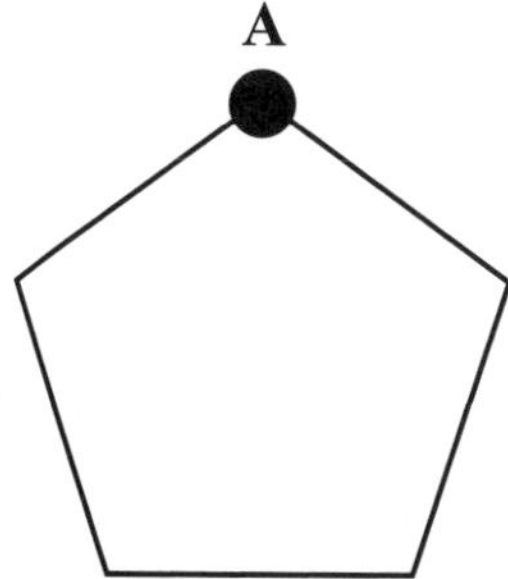

For Problems 8–10, evaluate if $a = 4$, $b = -5$, and $c = 2$.

8. $-b =$

9. $\dfrac{ab}{c} =$

10. $a + bc =$

   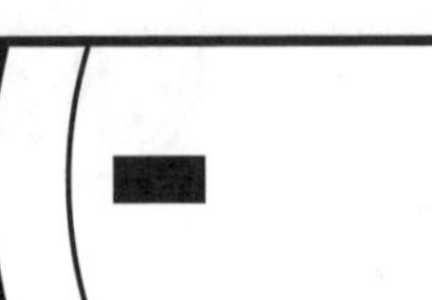 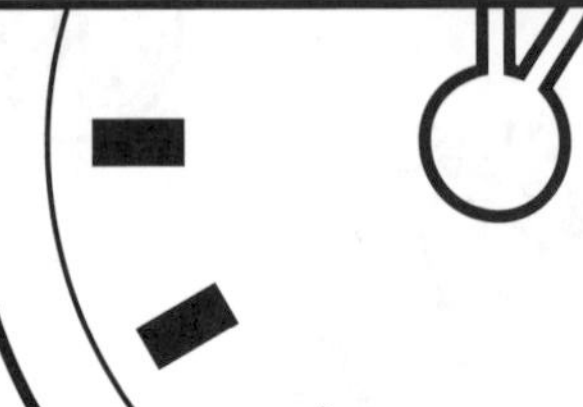

MINUTE 59

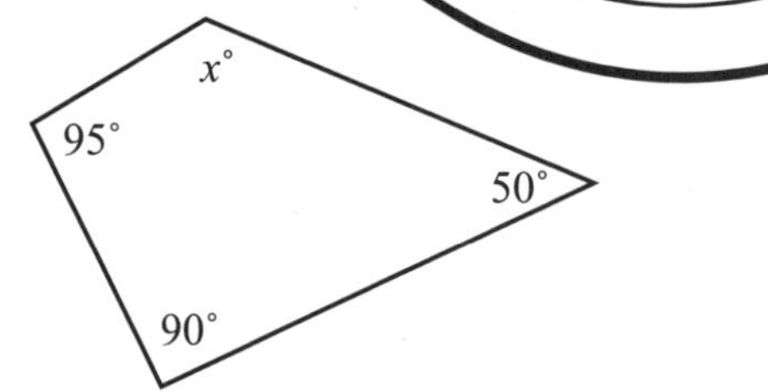

1. If the angles of a four-sided shape total 360°, then angle x is ________.

2. Circle the numbers that are greater than 2, but less than 2.4.

   2.03   2.41   1.99   2.22   3.1

3. The only even prime number is ________.

4. 16 weeks, 2 days is the same as ________.
   a. 105 days   **b.** 126 days   **c.** 114 days   **d.** 88 days

5. Leah is dealing cards. She deals a king, then a queen, then a king. The next card to be dealt will be:
   a. queen   **b.** king   **c.** can't tell   **d.** ace

6. What is the pattern in this sequence? _______________________

  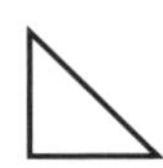

7. What is the lowest composite number with the factors of 2, 3, and 4? ________

8. Friends were sharing a bag of candy. Mike ate one-fourth of the candy. Shelby ate one-eighth of the candy originally in the bag. Then Shelby's dog ate one-half of the candy originally in the bag. How much candy remains? ________

For Problems 9–10, use the graph to the right.

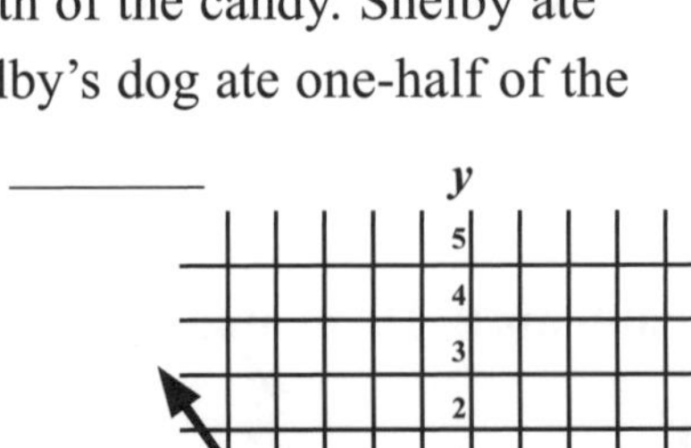

9. Where does the line cross the y–axis (y–intercept)? ________

10. What is the x-intercept? ________

MINUTE 60

1. You would most likely measure the width of a swimming pool in:

   a. cm   **b.** m   **c.** mm   **d.** km

2. Write the smallest possible number using the digits 4, 2, 8, 9, and 1. ___________

3. Do the shaded shapes to the right have the same perimeter?   Circle: Yes   or   No

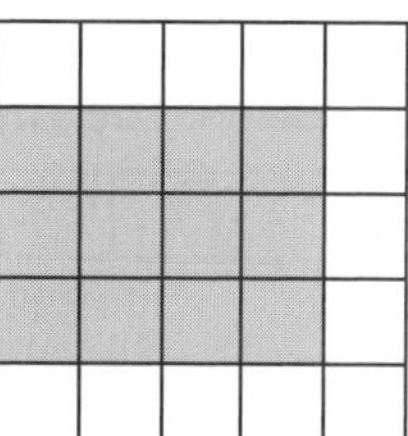 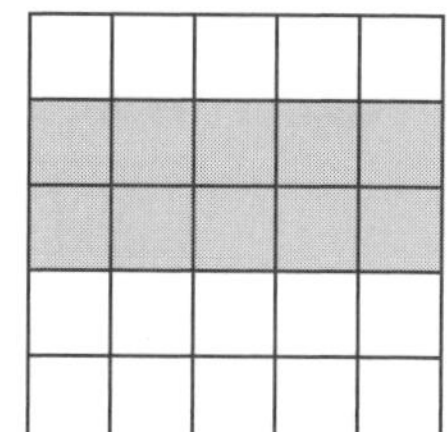

4. $(-8)^2 - 5 =$

5. Which shape below shows an obtuse angle? ___________

   a.   **b.**   **c.**

6. Complete the sequence: 4.8, 5.4, 6.0, ________, ________.

7. Circle three numbers below that have a sum of 7.

   -6   3   5   0   8

For Problems 8–10, use the graph to the right.

8. Which day of the week was the warmest? ___________

9. Which day of the week had the narrowest gap between the high and low temperatures? ___________

10. Which of these would be closest to the mean high temperature for the week?

   a. 90°   **b.** 40°

   c. 70°   **d.** 80°

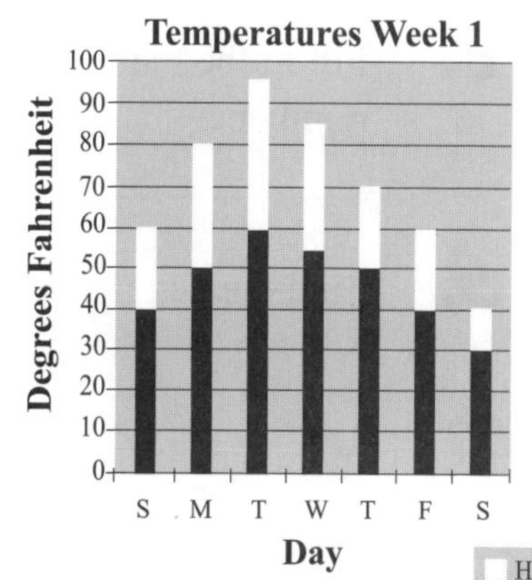

MINUTE 61

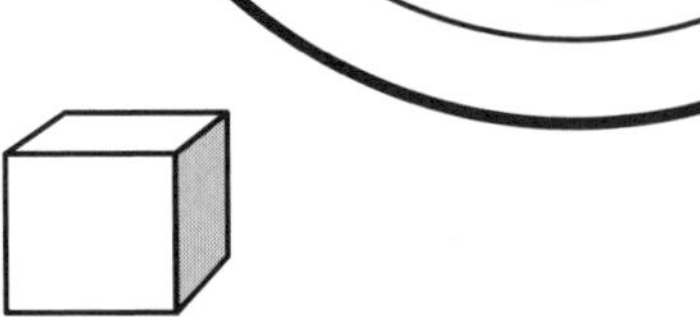

1. If the area of one side of this cube is 25cm^2, what is the area of the whole surface of the cube? _______

2. Fill in the missing number: $3 \cdot \boxed{} = 1.8$

3. What is the sum of the first four composite numbers in the list below? _______________

1	2	3	4	5	6	7	8	9	10

4. $-5 + -7 + 10 + 10 =$

5. If $-3(4 + a) = -15$, then $a =$ _______.

6. The length of each side of shape A has been doubled to create shape B. This means that the area of shape B is ______.

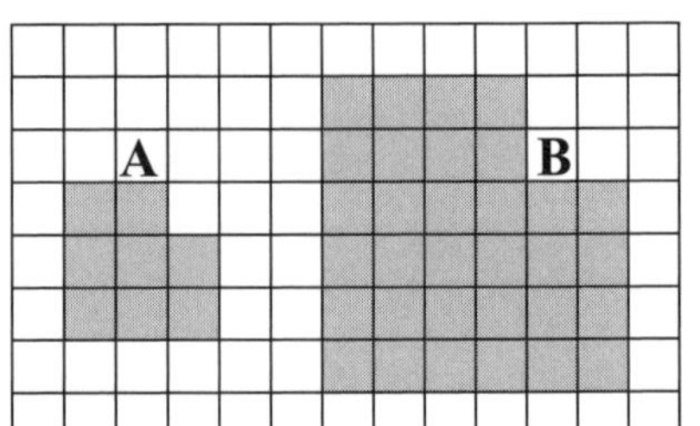

   a. doubled   **b.** three times bigger

   c. four times bigger   **d.** six times bigger

7. A number is between 20 and 30 and is three times the sum of its digits. What is the number? _______

8. Fill in the blanks using the numbers 7, 6, 2, 9, and 8 to make the smallest possible number.

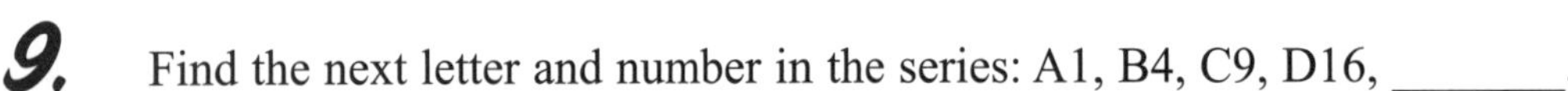

9. Find the next letter and number in the series: A1, B4, C9, D16, _______.

10. In the quadrilateral to the right, angle x equals _______.

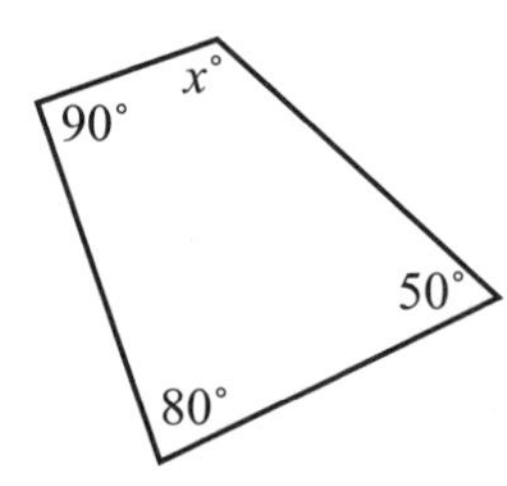

MINUTE 62

1. Add the two shaded areas together. (**Hint:** Each set of shaded and unshaded boxes represents a fraction. Find the sum.)

 +  =

For Problems 2–4, use the diagram to the right.

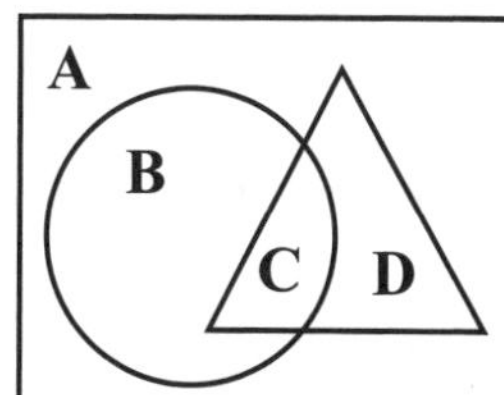

2. Which letter is inside the circle and the triangle? _______

3. Which letter is outside the circle but inside the triangle? _______

4. Which letter is outside the circle and the triangle? _______

x	y
1	-1
2	-4
3	-7

5. Look at the chart to the right and complete the function rule. $y = -3x +$ _______

6. Using the chart in Problem 5, if $x = 12$, then $y =$ _______.

7. Tom has four dollars. Bob has three times as much as Tom. Cindy has twice as much as Bob. How much do they have altogether? _________________

8. $\dfrac{4 + (-3)(-2)}{-2} =$

9. Circle the number that is different from the others.

4   6   7   9   12   15

10. Complete the bottom row of numbers on this chart.

			1			
		1	3	1		
	1	3	5	3	1	
1	3	5	7	5	3	1

MINUTE 63

1. Which shape below shows an acute angle? _______

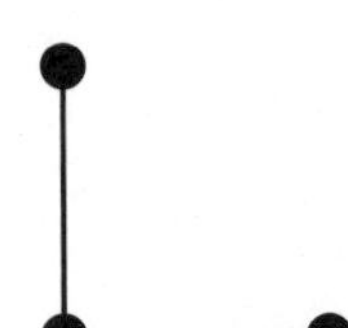
a.

b.

c.

2. An unknown number is half the product of 4 and 12. The number is _______ .

3. Jim's father is older than 40 but younger than 50. If you divide his age by 2, 4, 5, 8, or 10, there will be a remainder of 1. How old is Jim's father? _______

For Problems 4–6, use the coordinate grid to the right.

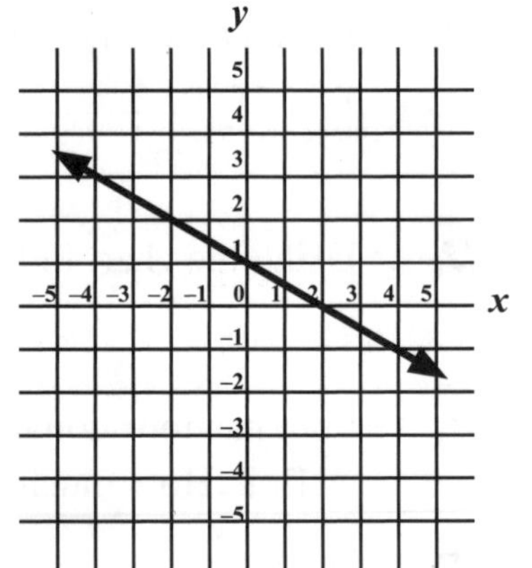

4. What is the y–intercept? _______

5. What is the x–intercept? _______

6. Does the line slope up or down? _______________

7. Find the dimensions of this rectangle.
Length = _______.
Width = _______.

> **Perimeter = 20 m**
> **Area = 21 m^2**

8. If pens cost 15 cents, how many can you buy with $3.00? _______

9. If one side of a cube has an area of 10 m^2, what is the surface area of the entire cube? _______

10. $4 + 3 \cdot (-2) =$

MINUTE 64

For Problems 1–5, match each word with its correct definition.

1. congruent   **a.** The amount of square units covering the outside of a shape.

2. similar   **b.** A triangle with two equal sides.

3. equilateral   **c.** Two figures with the exact same size and shape.

4. isosceles   **d.** Two figures with the same shape but different size.

5. surface area   **e.** A triangle with three equal sides.

6. Which number is three places to the right of the median? _______

1	2	3	4	5	6	7	8	9

7. Circle the numbers in the set $\{2, 3, 4, 5, 6, 7\}$ that make the inequality $3a + 13 > 14$ true.

2   3   4   5   6   7

8. $\left[\dfrac{3}{7}\right]\left[\dfrac{2}{3}\right] =$

9. $\dfrac{3}{11} \div \dfrac{2}{7} =$

10. Complete the chart if $y = 2x + 6$

x	y
-2	
	4
0	

MINUTE 65

1. Complete the times table.

×	7	8
-4		-32
-6	-42	

2. Write an equation that represents this statement: two times a number plus 1 is 11.

____

3. What number would solve the equation in Problem 2? _______

For Problems 4–6, cross out the item that does NOT belong on the list.

4.   5   9   16   100

5. $\dfrac{4}{8}$ $\dfrac{9}{18}$ $\dfrac{14}{28}$ $\dfrac{7}{12}$

6.   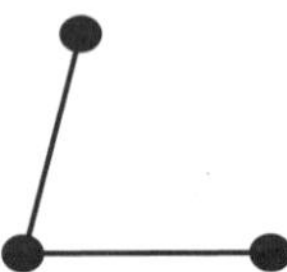         

For Problems 7–10, match the problems with their correct answers.

7. $13a = -26$ **a.** $a = 1$

8. $\dfrac{a}{4} = -5$ **b.** $a = -2$

9. $a - 11 = -10$ **c.** $a = -20$

10. $a + 3 = -14$ **d.** $a = -17$

MINUTE 66

1. Which letter is inside all three shapes? _______

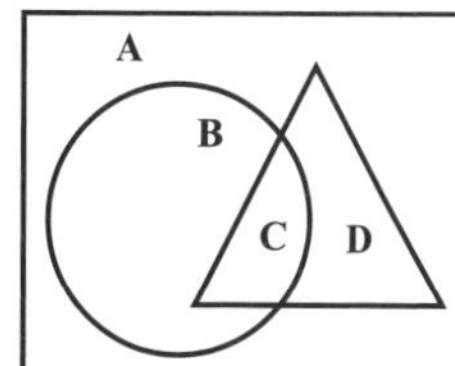

2. Which letter is inside the triangle but outside the circle? _______

3. Which of these shaded shapes has a perimeter of 14 units? _______

a. **b.** **c.**

4. Which shape in Problem 3 has the greatest area? _______

5. A shape with the greatest perimeter always has the greatest area.
Circle: True or False

Dog and Cat Owners

6. According to this Venn diagram, how many people have a dog? _______

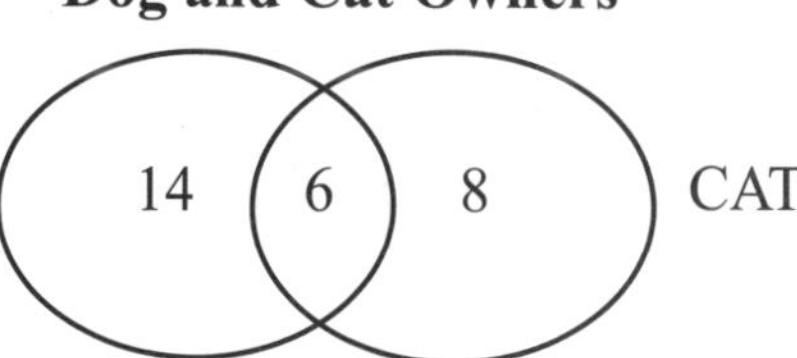

7. Complete the chart.

Fraction	Decimal	Percent
	0.2	

For Problems 8–10, use >, <, or = and let $a = -2$, $b = -4$, and $c = 5$.

8. ab _______ c

9. a^2 _______ $-b$

10. $\dfrac{1}{2}ab$ _______ $\dfrac{c}{0.5}$

MINUTE 67

1. What fraction of the total square is shaded? _____________

2. $\dfrac{1}{4} \cdot 24 =$

3. Complete this division table.

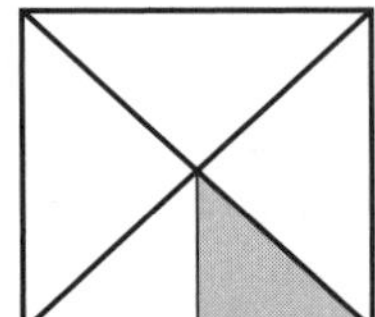

$\div$	12	18
-2		-9
-3	-4	

4. 20% of 70 =

5. Which shape below shows a right angle?

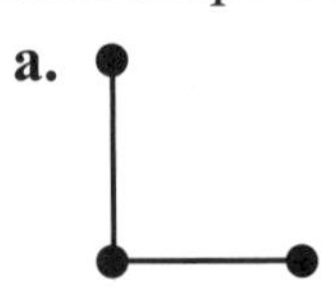 a.   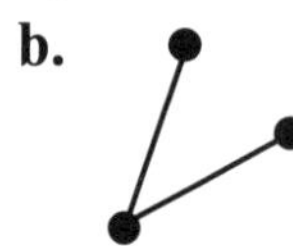 b.    c.

6. $2^3 - 5 =$

7. A is to ∀ as △ is to:

a. ▽   b. ▷   c. ◁

For Problems 8–10, use the graph to the right.

8. At what time did Jen finish her trip? _________

9. How many miles did Jen ride? _________

10. At what two times did Jen appear to take a break?
________ and ________.

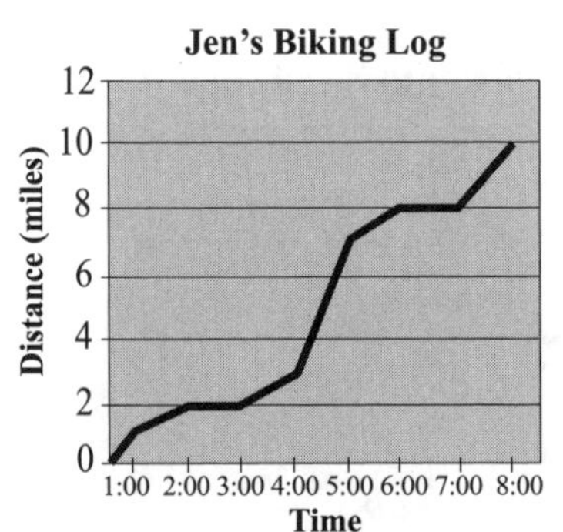

MINUTE 68

1. Fill in the remaining boxes to complete the pattern.

7			28	35		49

2. How many small cubes placed on top of the grid, fitting exactly on the squares, would it take to make a large cube? _______

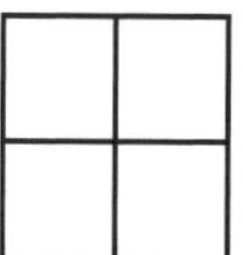

3. If $\dfrac{1}{4} - \dfrac{2}{3} + \dfrac{3}{5} = \dfrac{a}{60}$, then $a =$ _______.

4. Circle the numbers in the set {3, 6, 9, 12, 15} that make the inequality $\dfrac{a}{3} + 1^3 \geq 4$ true.

   3   6   9   12   15

For Problems 5–7, use the coordinate grid to the right.

5. The Roman numerals identify the quadrants. In which quadrant is point A? ___________

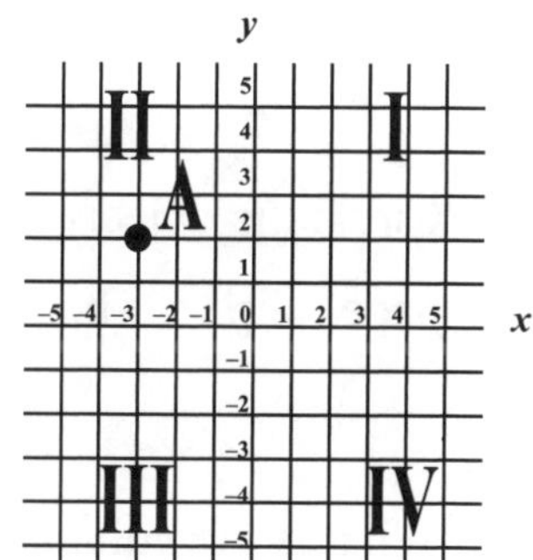

6. What are the coordinates of point A? ___________

7. In which quadrant would (5, -3) be? ___________

For Problems 8–9, use the chart to the right.

8. If the dot (B2) is shifted two squares south and two squares east, in which square will it be? _______

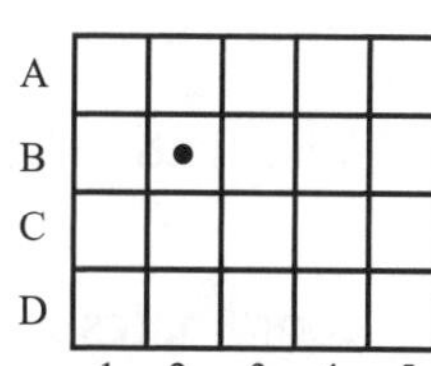

9. If the dot (B2) is moved one square northwest, in which square will it be? _______

10. Draw a vertical line of symmetry through the heart.

Minute 69

1. Complete this addition table.

+	-5	-6
3	-2	
8		2

2. Circle the numbers that can be divided evenly by 3, 4, and 5.

12   15   24   30   60

3. How many times bigger is the underlined 5 than the other 5 in the number 4$\underline{5}$,245?
a. 1,000 times   **b.** 100 times   **c.** 10 times

4. Circle the objects below that are longer than 1 meter.

calculator   mouse   bed   basketball   dining table

5. Circle the objects that are shorter than 5 centimeters.

paper clip   book   writing paper   pencil eraser   bottle cap

6. What is the volume of a box that is 6 in. × 8 in. × $\frac{1}{2}$ in.? _______

For Problems 7–10, match each word with its correct definition.

7. consecutive numbers   **a.** when numbers are in order from least to greatest

8. coordinates   **b.** numbers used to locate points on a grid

9. descending order   **c.** numbers that follow in order and are not interrupted

10. ascending order   **d.** when numbers are in order from greatest to least

MINUTE 70

1. What relationship do the arrows represent in the diagram?

____

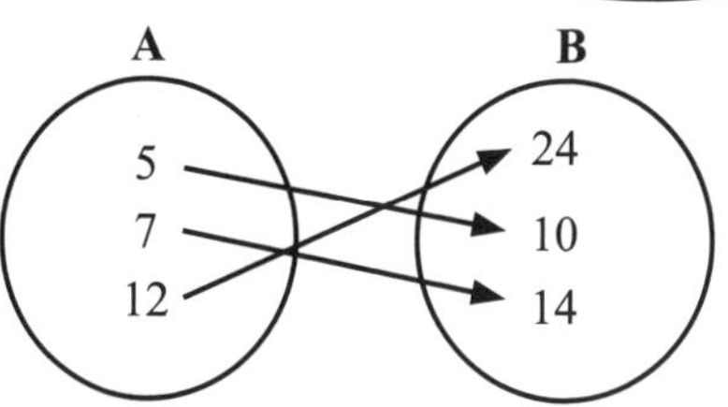

2. What fraction of the total shape is shaded? _______

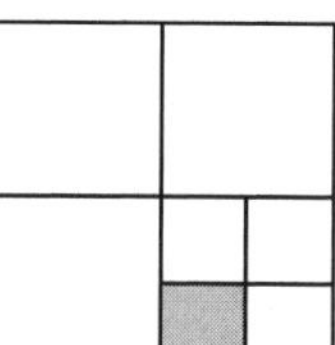

3. If 3! = 3 • 2 • 1, what does 4! equal?
  a. 6   **b.** 12   **c.** 24   **d.** 120

4. Which of these is an equilateral triangle? _______

a.   **b.**   **c.**   **d.**

5. Which shape in Problem 4 is a right triangle? _____________

For Problems 6–7, use the pie chart to the right.

6. Shade 25% of the pie chart.

7. If six slices of the pie chart were shaded, what percent would that represent? _________

For Problems 8–10, use the graph to the right.

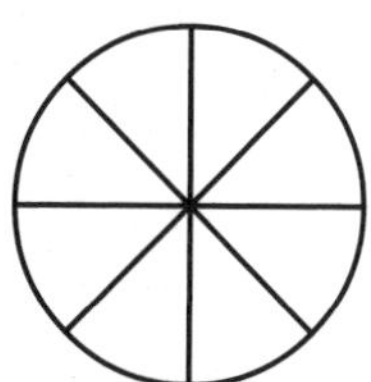

8. In which quadrant would the point (3, 3) be? _______

9. In which quadrant would the point (-2, -5) be? _______

10. Does the line have a positive slope or a negative slope? _____________

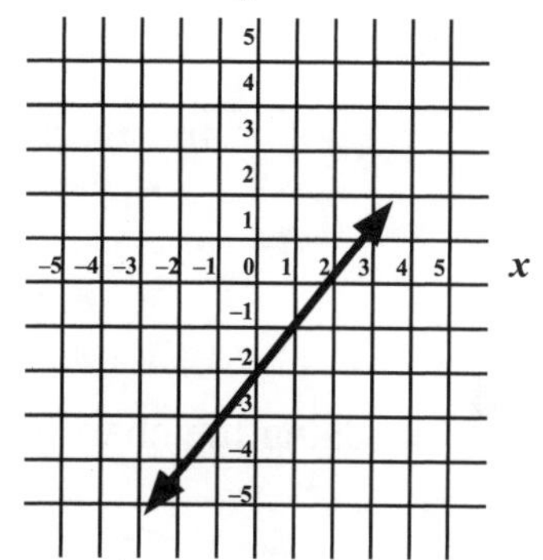

MINUTE 71

For Problems 1–3, use the stem-leaf plot to the right.

1	1 2 2
2	2 6 8
3	0 1 2
5	5 5 5 6
6	1 3 5
7	2 3
9	4 6

KEY
6|1 represents 61

1. What number is the mode of the plot? _______

2. Does the number 64 appear on the plot? _______

3. How many numbers are represented by the plot? _______

For Problems 4–7, use the spinner diagram to the right.

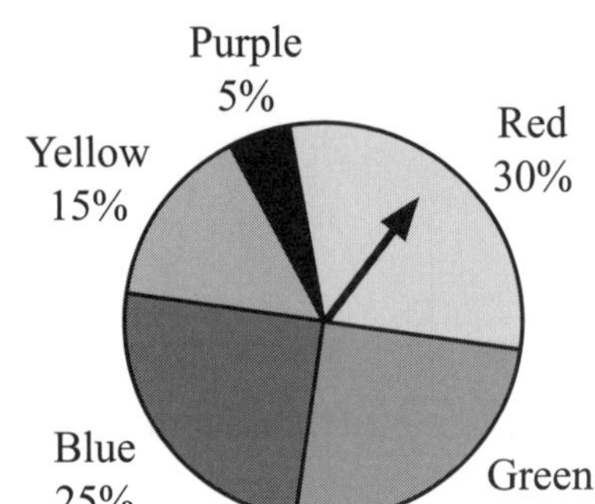

4. On which color is the spinner most likely to stop? _______________

5. Is there a better chance of spinning Blue or Yellow? _____________

6. If the spinner is spun 100 times, what is the average number of times it would stop on Red? _______

7. The spinner will land on Blue or Green about half the time on average.
Circle: True or False

8. $-3 + \dfrac{-12}{-2} =$

9. Look at the chart to the right and write the function rule.

$y =$ _______

x	y
1	3
2	6
3	9

10. Using the chart in Problem 9, if $x = -3$, then $y =$ _______.

MINUTE 72

For Problems 1–3, use the stem-leaf plot to the right.

1	1 1 5 7
2	2 2 2 4
3	0 6 8 9
4	3 4 5 6
5	2

KEY
4|3 represents 4.3

1. How many times does the number 2.2 show up? _______

2. How many numbers are between 4.5 and 5.0? _______

3. What is the range (biggest number–smallest number) of the plot? _______

4. Complete this subtraction table.

−	-5	-6
3	-8	
8		-14

5. Which of these fractions is closest to zero? _______

    a. $\dfrac{1}{8}$ **b.** $\dfrac{1}{10}$ **c.** $\dfrac{2}{50}$ **d.** $\dfrac{9}{10}$

6. Which of these shapes has the most sides? _______

    a. decagon   **b.** octagon   **c.** pentagon   **d.** hexagon

For Problems 7–10, use the clues to complete the crossword.

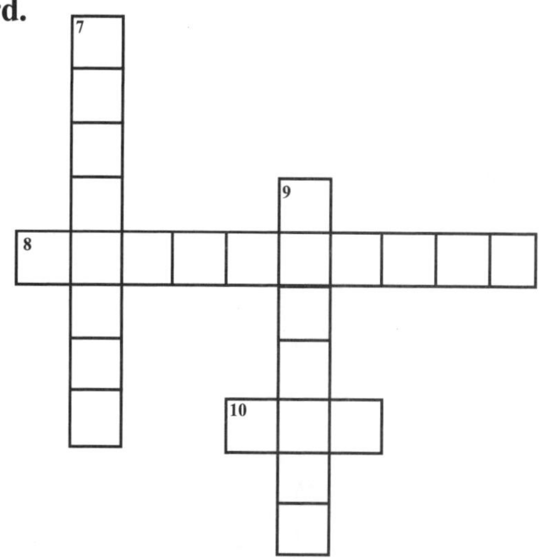

7. The answer to a division problem.

8. The answer to a subtraction problem.

9. The answer to a multiplication problem.

10. The answer to an addition problem.

MINUTE 73

1. Complete this factor tree.

2. Use •, +, −, or ÷ to complete. $3 \;\square\; 12 \;\square\; 4 = 6$

3. If $y + 1.7 = 1$, then $y =$ _______.

4. If $d = 3$, does $d + d + d = 3d$?   Circle: Yes   or   No

5. Complete this multiplication table.

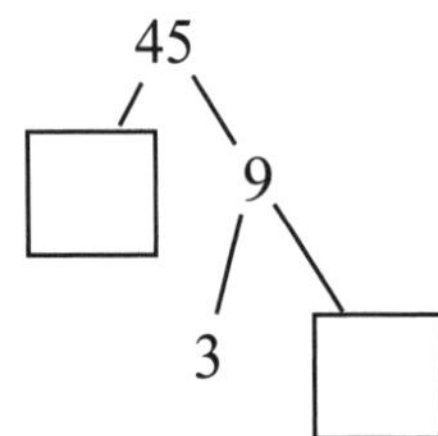

×	-5	-6
3	-15	
8		-48

6. If $\pi = 3.14$, then $10\pi =$ _______.

For Problems 7–10, match each expression with an equivalent expression.

7. $a \bullet a \bullet a$ **a.** $\dfrac{a}{3}$

8. $a + a + a$ **b.** $-a$

9. $a \div 3$ **c.** $3a$

10. $a - a - a$ **d.** a^3

MINUTE 74

1. Put the numbers 23, 35, 26, 38, and 39 into the stem-leaf plot to the right.

2			
3			

2. What is the median number in Problem 1? _______

3. Fill in the missing number in the box.

$3 \Big\langle$

$6 \rightarrow 9 \rightarrow 12 \rightarrow 15$

$6 \rightarrow 12 \rightarrow 24 \rightarrow \boxed{}$

4. The numbers in the boxes are all multiples of 4 that are less than 40. Fill in the missing number.

4	36	16
12		32
28	8	20

5. What is the sum of row 1 in the chart in Problem 4? _______

6. If the time is 4:40, what time was it 70 minutes ago? _______

For Problems 7–10, use the clues to complete the crossword.

7. The number in the middle of an ordered group.

8. An angle that is less than 90 degrees.

9. The number in a group that shows up the most often.

10. The largest number in a group minus the smallest.

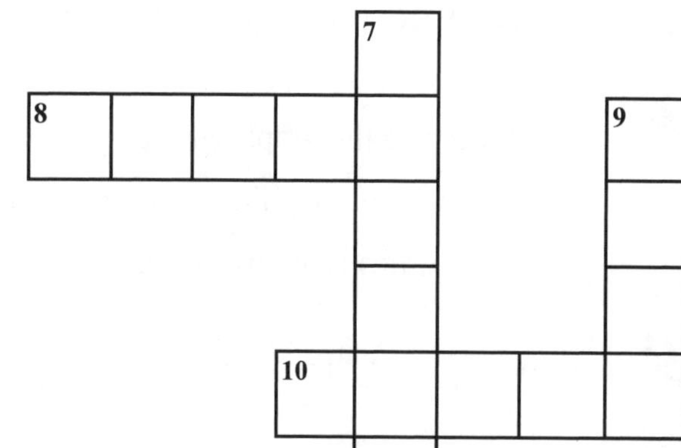

MINUTE 75

1. Write in the simplest form: $\dfrac{16}{20}$ =

2. Estimate: $42 \times 58 \approx$ _______. (**Hint:** "≈" means "approximately")

3. What number times 7 equals negative 56? _______

4. How many dimes are in $6.00? _______

5. Complete this addition table.

+	-4	-5
-6	-10	
-7		-12

6. How many cookies are in 3.5 dozen? _______

7. The distance around a circle is sometimes referred to as _______.

a. diameter   **b.** radius   **c.** circumference   **d.** pi

For Problems 8–10, use the graph to the right.

8. According to the graph, group _______ has twice as many points as group D and _______ times as many points as group B.

9. Group _______ has half as many points as group E.

10. Altogether, groups A, B, and C have a total of _______ points.

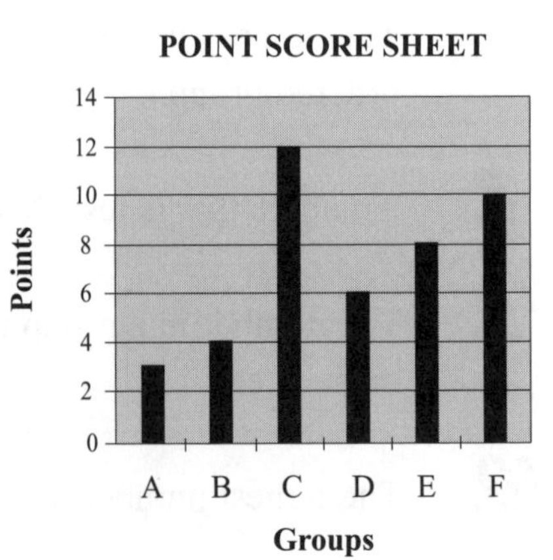

MINUTE 76

1. How many fourths are in $5\frac{1}{2}$? _______

2. If four apples cost $0.40, how much would six apples cost? _______

3. If a triangle has a 50 degree angle and a 60 degree angle, how many degrees is the third angle? _______

4. A $30 shirt is 50% off. What is the new price? _______

5. What is your change from a $20 bill if your dinner costs $11.80? _______

For Problems 6–8, use the stem-leaf plot to the right.

Bowling Scores

10	5 6
11	2 2
12	1 1 7
13	0 5 5
14	2 6
15	1 3 3
16	0 4 4 4
17	2 5 5 7
18	6 6 7
19	2 8
20	5

KEY
15|1 represents 151

6. What was the highest score recorded? _______

7. What was the lowest score recorded? _______

8. What was the mode score? _______

9. Fill in the missing numbers to complete the pattern.

2	7	2	11	2	15	2		

10. What is the radius of this circle? _______

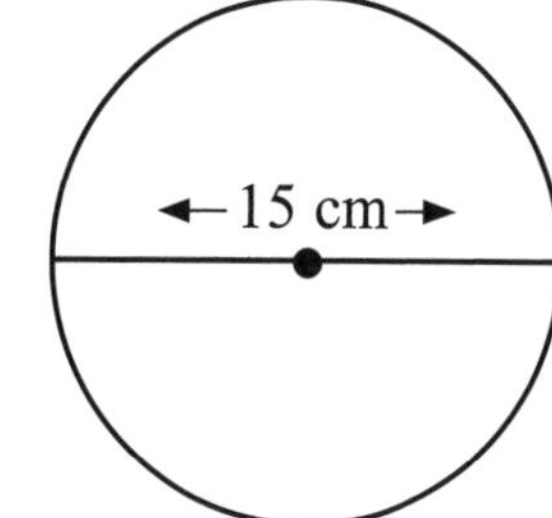

MINUTE 77

For Problems 1–2, use the grid to the right.

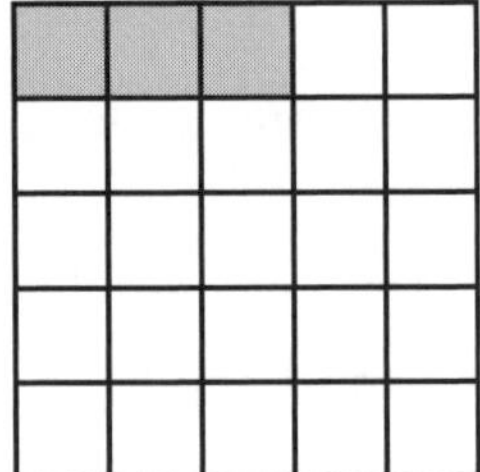

1. If two more of the squares were shaded, what total percent would be shaded? _______

2. How many small cubes placed on top of the grid, fitting exactly on the squares, would it take to make a large cube? _______

3. $5^2 - 33 =$

4. How many thirds are in 7? _______

5. What is the perimeter of a 5 in. × 9 in. picture frame? _______

6. Would a 40 in.^{2} picture fill a 5 in. × 9 in. picture frame?   Circle: Yes   or   No

For Problems 7–10, match each statement with its correct algebraic expression.

7. three more than a number squared   **a.** $\dfrac{1}{3}n$

8. three less than twice a number   **b.** $\dfrac{n^3}{3}$

9. a number cubed divided by 3   **c.** $n^2 + 3$

10. one-third of a number   **d.** $2n - 3$

MINUTE 78

For Problems 1–3, use the diagram to the right.

1. Which letter is inside the pentagon and the octagon? _______

2. Which letter is inside the octagon and the oval? _______

3. Which letter is outside the octagon and the pentagon? _______

4. Bananas cost 50 cents each and oranges cost 75 cents each. How much will two of each cost? _________________

5. What is the mean of 30 and 50? _______

6. A $40 jacket is 25% off. How much will you save? _________________

For Problems 7–8, use the table to the right.

7. What is the sum of the numbers in row B? _______

8. What is the product of the numbers in row A? _______

A	-3	5	-4
B	-2	-6	-8
C	-1	0	7

For Problems 9–10, use the graph to the right.

9. In what quadrant is point A located? _______

10. What is the y-intercept of the line? _______

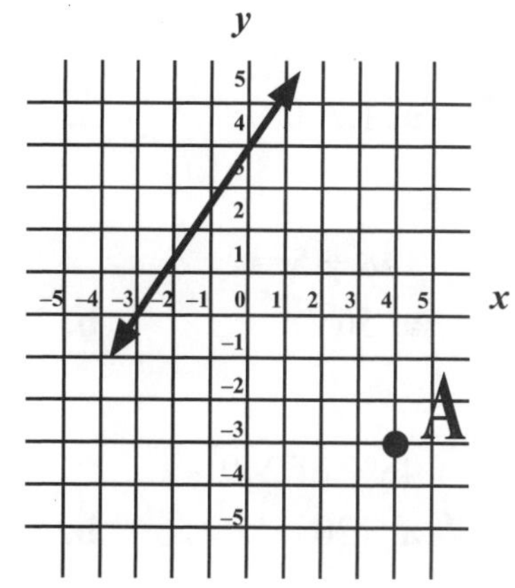

MINUTE 79

For Problems 1–3, use the Venn diagram to the right.

1. What number is in all three circles? _______

2. Which number(s) are in both circles A and B? ___________

3. How many different numbers are in circles A and C? _______

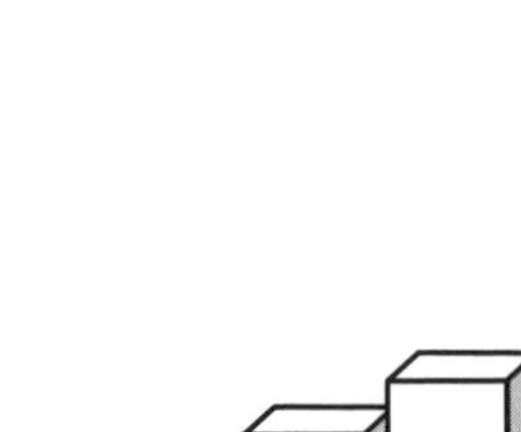

4. What is the interest for one year at 10% on $2,500? _______

5. Complete this subtraction table.

$-$	-4	-5
-6	2	
7		-12

For Problems 6–7, use the picture to the right.

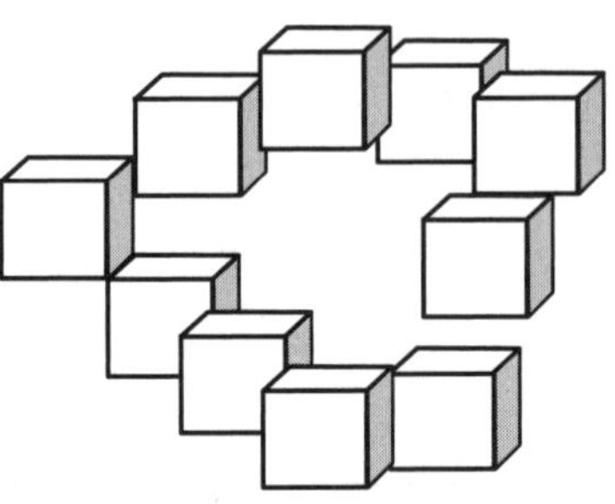

6. How many cubes are in the picture? _______

7. If each cube has six faces, how many total faces are in this picture? _______

For Problems 8–10, estimate to find the best answer.

8. $82 \times 41 =$
  a. 1,200   **b.** 120   **c.** 1,600   **d.** 3,200

9. $148 \div 5 =$
  a. 50   **b.** 30   **c.** 75   **d.** 25

10. 48% of 240 =
  a. 120   **b.** 250   **c.** 24   **d.** 160

MINUTE 80

1. If it is 10:46 a.m., how many minutes until noon? _______

2. Mark has a string that is six feet long. If he cuts it in half and then cuts each half in half, how long will each piece be? _______

For Problems 3–6, circle three items that are of equal value.

3. 3^3 3×3 27   9 $3 \cdot 3 \cdot 3$

4. 100 $\dfrac{1{,}000}{10}$ 10 10^2 $10 \cdot 10 \cdot 10$

5. Hexagon  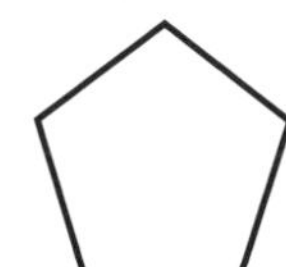 Pentagon 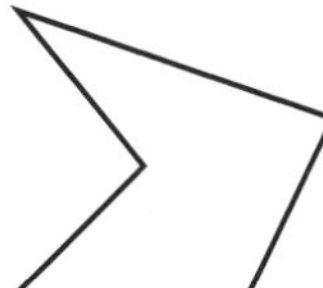

6. 1.265×10^2 0.1265×10^3 12.65×10^1 1.265×10^3 0.1265×10^4

7. If Jerome walks 1.6 miles to school each day, how long is the round-trip? _______

8. How many legs do six chickens and four cows have in total? _______

9. Fill in the missing squared numbers.

1	4	9	16		

10. NET is to TEN as 304 is to:
   a. 340   **b.** 430   **c.** 403   **d.** 304

MINUTE 81

1. $0.25 + 50\% \ \dfrac{1}{10} =$

2. Using the numbers 2, 6, 5, 1, and 8, fill in the lines below to create the greatest number possible.

______ ______ . ______ ______ ______

For Problems 3–5, use > , < , or = .

3. $\sqrt{36}$ ______ -8

4. $0.4\overline{6}$ ______ 0.48

5. Obtuse Angle ______ Acute Angle

6. The letter **M** has two ______ lines.   Circle: Parallel   or   Perpendicular

For Problems 7–10, fill in the boxes to complete the correct math equations.

<table>
<tr><td>7</td><td></td><td></td><td></td><td></td><td></td></tr>
<tr><td>÷</td><td></td><td></td><td></td><td></td><td></td></tr>
<tr><td>-5</td><td>8</td><td>-4</td><td></td><td></td><td></td></tr>
<tr><td>=</td><td></td><td>•</td><td></td><td></td><td></td></tr>
<tr><td>9</td><td>-9</td><td>÷</td><td>2</td><td>•</td><td>6</td><td>=</td><td></td></tr>
<tr><td></td><td></td><td>=</td><td></td><td></td><td></td></tr>
<tr><td></td><td>10</td><td></td><td>+</td><td>-7</td><td>=</td><td></td></tr>
</table>

MINUTE 82

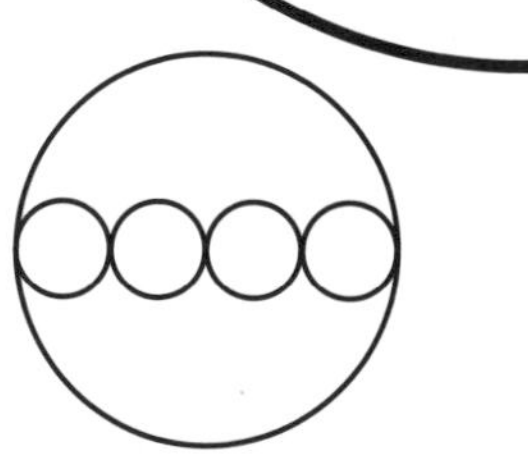

1. If the diameter of the largest circle is 40, what is the diameter of the smaller circles? ______________

(**Hint:** All small circles are congruent.)

For Problems 2–5, use the coordinate grid to the right.

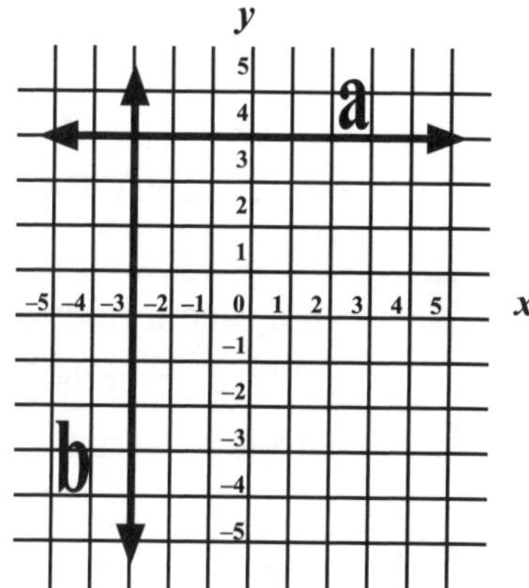

2. Line *a* and line *b* are ________.

Circle: parallel   or   perpendicular

3. Where do lines *a* and *b* intersect? ________

4. In which quadrant do the lines intersect? ________

5. If (x, y) is a point in Quadrant I, then $(-x, -y)$ is in :

Circle: Quadrant II   Quadrant III   Quadrant IV

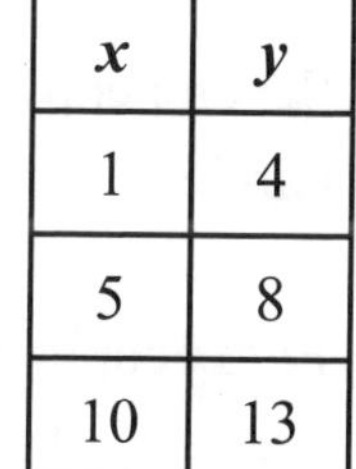

6. Look at the chart and write a function rule.

$y =$ ________

x	y
1	4
5	8
10	13

7. Which of the following would be the next term in this sequence?

Ab5, Cd7, Ef9, ________

a. GH11   **b.** Gh13   **c.** Gh11   **d.** GH13

For Problems 8–10, shade the box with the best equivalent fact.

8. 1 yard

24 inches	4 feet	36 inches

9. Degrees in a triangle

180	90	360

10. Degrees in a quadrilateral

180	90	360

Minute 83

1. If a snail moves six feet in 15 minutes, how far will it go in two hours? _______________

2. Use the digits 1, 6, and 7 to fill in the remaining squares so that no two consecutive numbers are beside each other vertically, horizontally, or diagonally.

	3	5	
		8	2
	4		

For Problems 3–6, use the Venn diagram to the right.

Television Survey

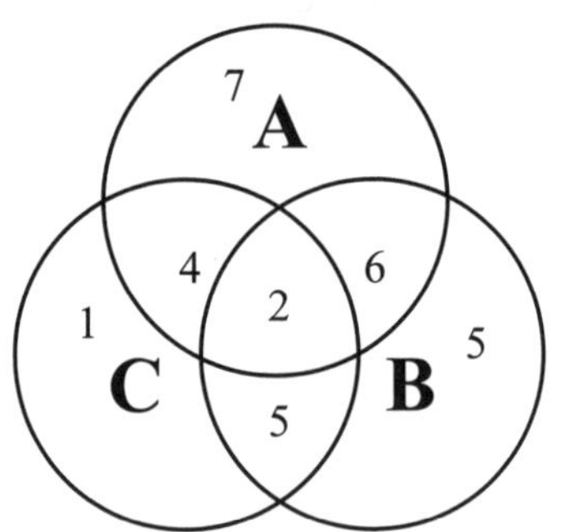

3. In a recent television survey, only two people preferred all three brands (A, B, C).
Circle: True   or   False

4. Eight people preferred brands A and B.
Circle: True   or   False

5. Seven people preferred brand A only.
Circle: True   or   False

6. Five people preferred brands C and B, but not brand A.
Circle: True   or   False

7. $20\% + \dfrac{2}{5} + 0.08 =$

8. $\sqrt{2^2 + 5} =$

9. $\sqrt{4} \cdot \sqrt{9} =$

10. I am an even number less than 30 but more than 20. I am also a multiple of 3. What number am I? _______

Seventh-Grade Math Minutes © 2007 Creative Teaching Press

MINUTE 84

For Problems 1–3, use the shape to the right.

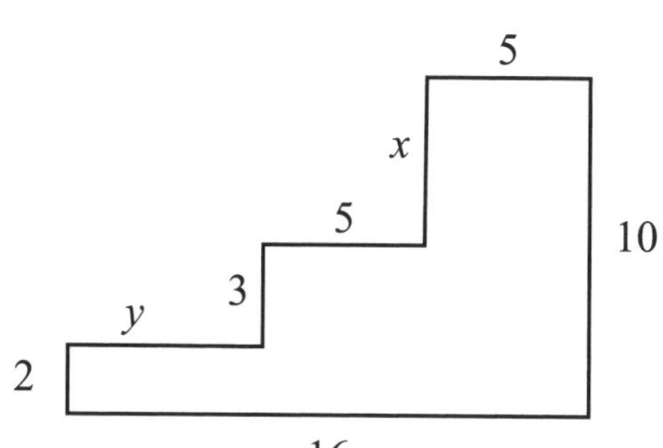

1. $x =$ ______

2. $y =$ ______

3. What is the perimeter of the shape? ______

4. If $\dfrac{6}{42} = \dfrac{n}{7}$, then $n =$ ______.

5.
$$
\begin{array}{r}
3 \text{ days} \quad 18 \text{ hours} \\
+\ 2 \text{ days} \quad\ 6 \text{ hours} \\
\hline
\end{array}
$$

6. $30\% + \dfrac{1}{5} + 0.12 =$

For Problems 7–10, use the following clues to complete the crossword.

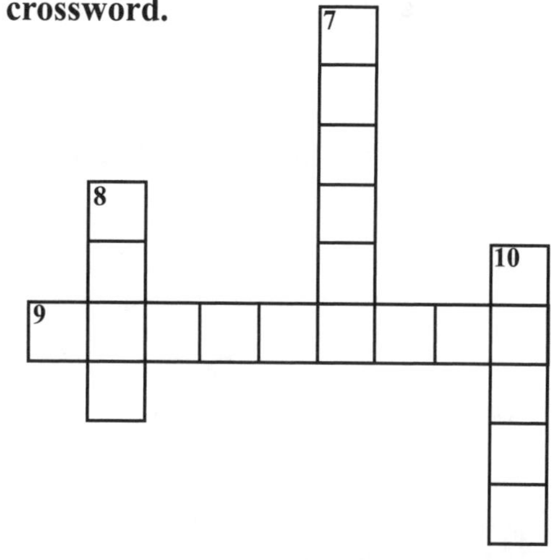

7. The amount of three-dimensional space taken up by an object.

8. The amount of square units inside a shape.

9. The distance around a shape.

10. A number that can only be divided by 1 and itself.

MINUTE 85

1. The letter **H** has _______ lines.
    a. parallel   **b.** perpendicular   **c.** both parallel and perpendicular

2. There are four aces in a deck of 52 cards. What are the chances of drawing an ace from a deck on one draw? _______

3. Write 7.25 as a fraction. _______

4. $5 + 5 \cdot 5 - 5 \div 5 =$

5. Ellen likes to draw pentagons and hexagons. Her paper has a total of 39 sides. If there are four hexagons, how many pentagons are there? _______

6. If $d - 3.6 = 7.4$, then $d =$ _______.

7. To turn 168 hours into days, you should _______.
    a. divide by 60   **b.** multiply by 24   **c.** divide by 24   **d.** multiply by 7

For Problems 8–10, use the chart to the right.

8. 3 gal. = _______ qt.

9. 6 pt. = _______ qt.

10. 2 qt. = _______ oz.

1 gal. = 4 qt.
1 qt. = 2 pt.
1 pt. = 16 oz

MINUTE 86

1. If the time is 2:12 p.m., then how many minutes ago did the time turn to noon?

____

For Problems 2–3, use the picture to the right.

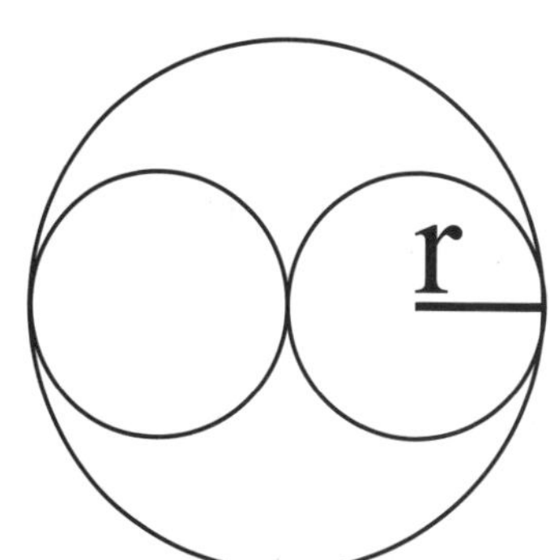

2. The diameter of the largest circle is 24. What is the radius of the smaller circles? _______

3. What is the diameter of the smaller circle? _______

For Problems 4–6, evaluate if $a = 3$ and $b = 4$.

4. $a^2 + b^2 =$

5. $\sqrt{a^2 + b^2} =$

6. $(ab)^2 =$

For Problems 7–10, use the chart to the right.

x_1	x_2	y_1	y_2
2	-2	12	4

7. In which quadrant would the point (x_2, y_2) be? _______

8. Solve: $y_2 - y_1 =$

9. Solve: $x_2 - x_1 =$

10. Find the slope of the line that contains the points listed in the chart.

$$slope = \frac{y_2 - y_1}{x_2 - x_1} =$$

MINUTE 87

For Problems 1–3, use the grid to the right.

1. What is the sum of column A? _______

2. What is the product of column B? _______

3. What is the product of column C? _______

A	B	C
-2	-1	10
5	-4	0
-8	-6	-9

4. $75\% + \dfrac{1}{10} + 0.02 =$

5. $\sqrt{6^2 + 8^2} =$

6. Which of the following shapes would be next in the pattern?

a.      b.      c.   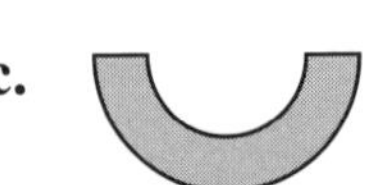   d.   

For Problems 7–10, use the coordinate grid to the right.

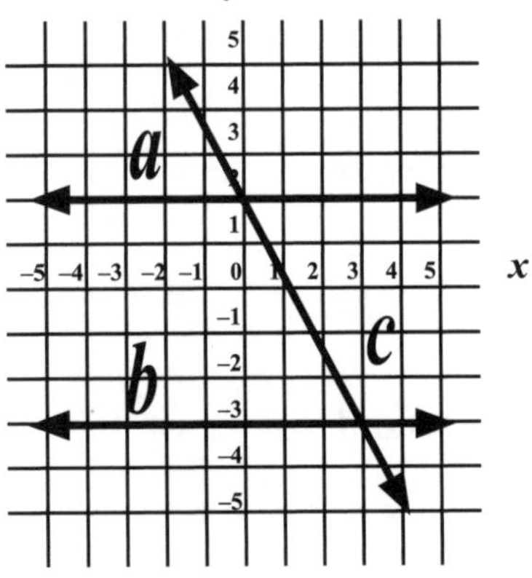

7. Lines a and b are _______.

Circle: parallel   or   perpendicular

8. Lines a and c intersect at (_______, _______).

9. Lines b and c intersect in Quadrant _______.

10. Line b has a y-intercept of _______.

MINUTE 88

1. To solve the equation $2x - 3 = 9$, you should first _______.

a. add 3   **b.** subtract 3   **c.** divide by 2   **d.** multiply by 2

2. The price of a \$40 jacket is marked down to \$30. What percent off is the jacket?

____

3. $\dfrac{5^4}{5^2} =$

4. In order to find 34% of 410, you should _______.

a. multiply 0.34 by 410   **b.** divide 0.34 by 410

c. multiply 0.034 by 410   **d.** divide 0.034 by 410

5. $2\sqrt{49} =$

6. Is the $\sqrt{11}$ closer to 3 or 4? _______

7. What is another way to write $a \cdot a \cdot a \cdot a$? _______

8. $\left[\dfrac{2}{5}\right]^2 =$

9. $\dfrac{4 \cdot 6 \cdot 7 \cdot 2}{6 \cdot 14} =$

10. If Rob made 7 out of 10 shots in a basketball game, what percent of shots did he miss? _______________

MINUTE 89

1. Write $3 \cdot 3 \cdot 3 \cdot 3 \cdot 5 \cdot 5$ using exponents: _______________

2. Which of the following is equal to $2^3 \cdot 2^2$?
   a. 2^6 **b.** 2^5 **c.** 2^1 **d.** 2^4

3. $2(4 + 1)^2 =$

4. $5(0.7 + 0.4) =$

5. Which value of n will make $4n > 22$ true?
   a. 4   **b.** 5   **c.** 6   **d.** -5

6. If $|-5| = 5$, then $|-12| =$ _______.

7. If $y = x^2$ and $x = 4$, then $y =$ _______.

8. Which of the following is the greatest number?
   a. 4^2 **b.** 2^4 **c.** $\dfrac{50}{2}$ **d.** three dozens

For Problems 9–10, use the Venn diagram to the right.

Food Survey

9. In a recent food survey, how many people preferred all three brands? _______

10. Seven people preferred brands _______ and _______.

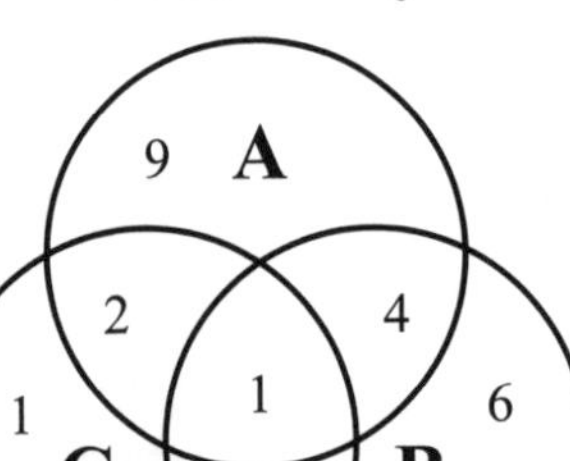

MINUTE 90

1. On Tuesday, Joe lost $10. On Wednesday, he made $5. On Thursday, he made $4. Did he make or lose money over those three days? _______________

2. $-3(-4 + -3) =$

3. Original price: $100   New Price: $72   What is the % decrease? _______

4. When dividing fractions, you must flip the _______ fraction over and then multiply the resulting fractions.   Circle: first   or   second

5. $\left[\dfrac{3}{8}\right]\left[-\dfrac{5}{7}\right] =$

6. Which one of the following is equal to 12%?
  a. $\dfrac{12}{100}$ **b.** $\dfrac{6}{50}$ **c.** 0.12   **d.** 0.012

7. Write using exponents: $4^2 \cdot 4 \cdot 4 \cdot 4 =$

8. $\left|-15\right| =$

9. How many halves are in 13? _______

10. What is the perimeter of this regular pentagon if each side is 1.3 inches? _______

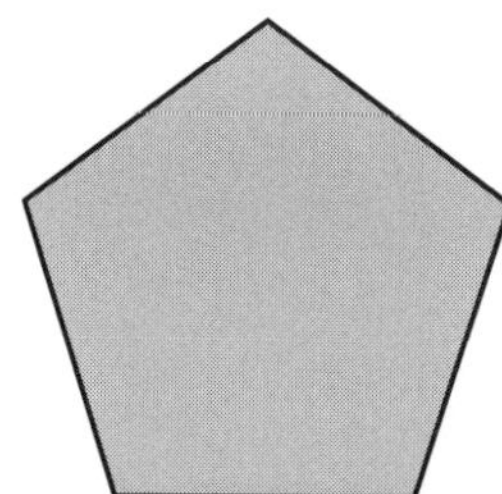

MINUTE 91

1. Which one of the following problems is incorrect?

   a. $-2 + -3 = -5$ **b.** $-2 \cdot -3 = -6$ **c.** $-8 \div -2 = 4$ **d.** $-4 - (-6) = 2$

2. $3|-5| =$

3. Reduce: $\dfrac{20}{50} =$

4. What percent is $\dfrac{20}{50}$? _______

5. Write as a decimal: $\dfrac{20}{50} =$

6. Which is greater, the mean or median of the numbers 1, 3, and 8? _____________

7. Write as an improper fraction: $3\dfrac{2}{7} =$

8. If $\dfrac{3}{4} = \dfrac{x}{36}$, then $x =$ _______.

9. Find three prime numbers whose product is 30. _______, _______, _______.

10. If all three angles of this triangle are equal, then $x =$ _______.

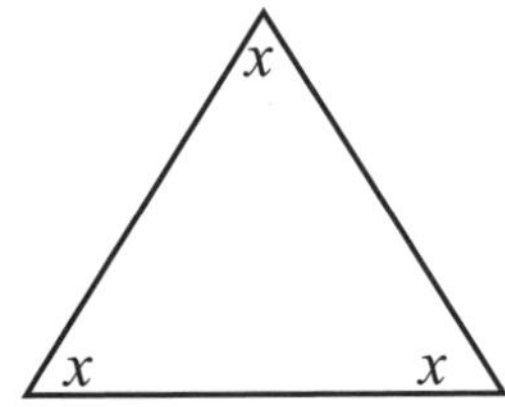

MINUTE 92

1. Fill in the missing number in the box.

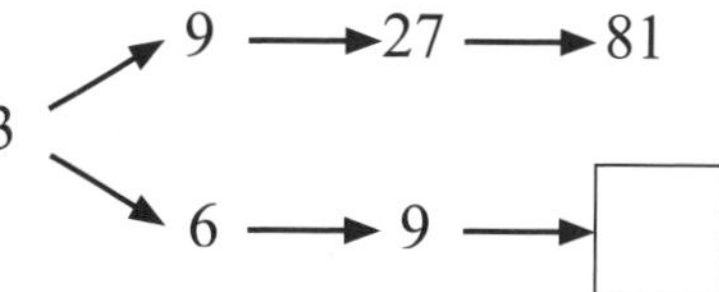

2. Complete this times table.

×	-4	-5
-6	24	
7		-35

3. Write as an improper fraction: $14\frac{1}{2} =$

For Problems 4–6, use the graph to the right.

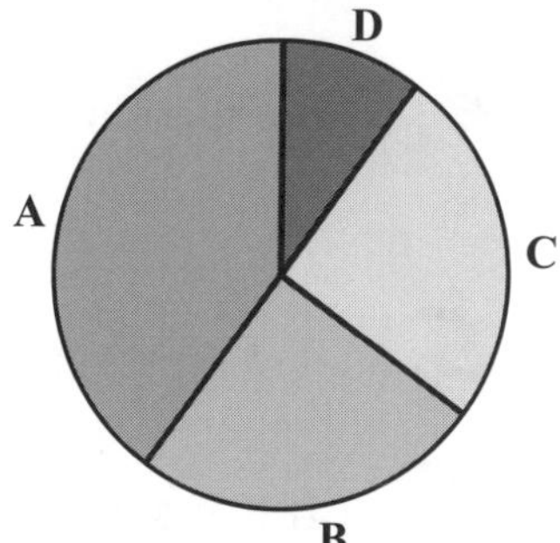

4. What percent of the graph does category B represent?
  a. 25%   **b.** 50%   **c.** 75%   **d.** 10%

5. If categories A, B, and C represent 90%, then category D represents _______.

6. Categories B and C appear to represent _______% of the graph.

7. $\dfrac{\frac{20}{2}}{5} =$

For Problems 8–10, use >, <, or =.

8. $\dfrac{4}{13}$ _______ $\dfrac{8}{26}$

9. $0.0\overline{2}$ _______ 0.02

10. $|\text{-}20|$ _______ $(\text{-}5)^2$

MINUTE 93

For Problems 1–3, use the chart to the right.

1. $y_2 - y_1 =$

x_1	x_2	y_1	y_2
2	-1	-3	6

2. $x_2 - x_1 =$

3. Find the slope of the line that contains the points from Problems 1 and 2.

4. Put these in order from least to greatest: -5, -7, $|{-5}|$, 0. __________________

5. $4^3 \bullet 4^8 =$

6. If two angles in a triangle are $60°$ and $100°$, is the third angle acute, obtuse, or right? ______________

7. $3(14 + 3 \bullet 12) =$

For Problems 8–10, use the coordinate grid to the right.

8. At what coordinates do the lines a and b intersect? ________

9. Lines b and c intersect in Quadrant ________.

10. Will line c intersect line a?   Circle: Yes   or   No

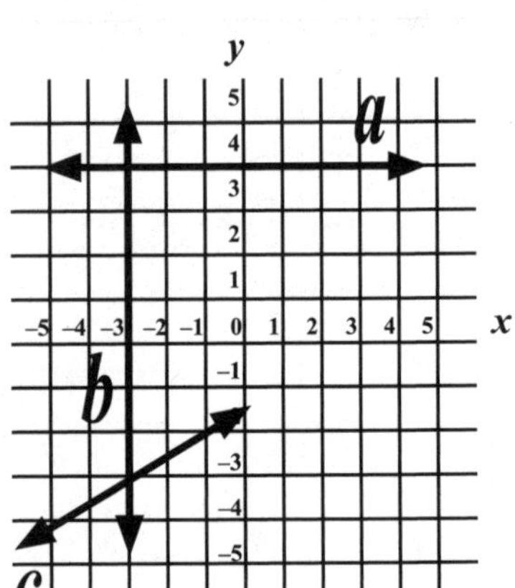

MINUTE 94

1. Which number is the length of this hypotenuse? _______   5

13

12

For Problems 2–4, use the parallelogram to the right.

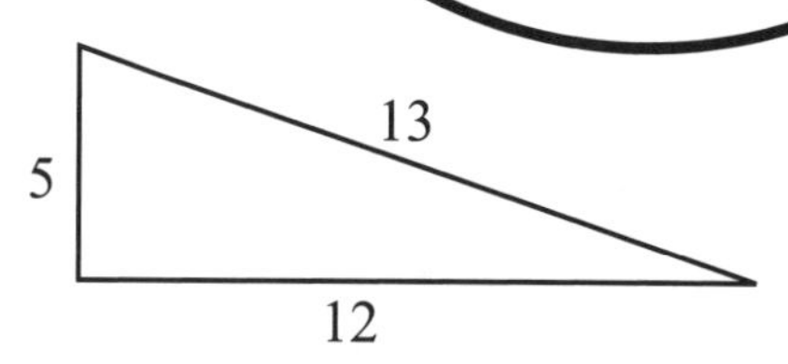
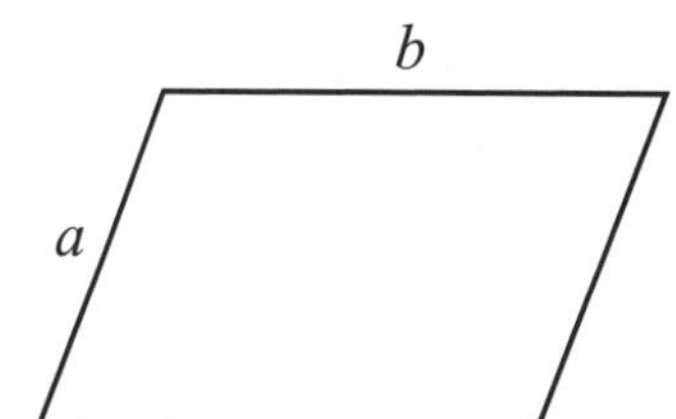

2. Using the letters on the parallelogram, what is the perimeter? _______

3. Using the letters on the parallelogram, what is the area? _______

4. If $a = 7$ and $b = 10$, the perimeter of the parallelogram is _______.

5. Fill in the missing numbers in this chart to complete the pattern.

4			16		24

For Problems 6–10, match each equation with its correct answer.

6. $3n = -63$ **a.** $n = -24$

7. $\dfrac{n}{-4} = 9$ **b.** $n = 6$

8. $2(n + 3) = 20$ **c.** $n = 7$

9. $0.5n = -12$ **d.** $n = -21$

10. $n^2 = 36$ **e.** $n = -36$

MINUTE 95

1. If Jenny's bill for her dinner is $32, how much should she leave for a 20% tip? _________

2. $\dfrac{1}{4} + 30\% + 0.02 =$

3. $\left[\dfrac{3}{9}\right]\left[-\dfrac{6}{3}\right] =$

4. $(-4) \cdot (-6) =$ _________ $(-7) \cdot (8) =$ _________ $(4) \cdot (-9) =$ _________

5. $\sqrt{\sqrt{16}} =$

For Problems 6–7, use the square to the right.

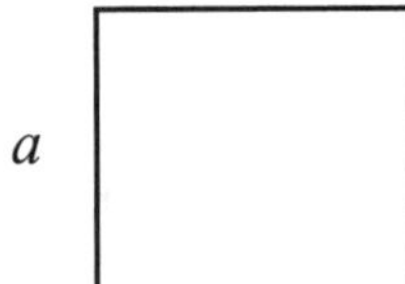

6. If the length of a side of the square is a units, what is its perimeter? ___________

7. What is the area of the square if $a = 7$ units? ___________

8. If $x = 2$, then $2x^2 - x =$ _______.

9. Use $y = 3x + 5$ to complete this chart.

x	y
-2	
5	
-4	

10. What four numbers are shown by this stem-leaf plot? _______, _______, _______, _______

1	3
2	5
3	6 8

KEY
1|5 = 15

MINUTE 96

1. Original price: $50   Final price: $60   The percent increase in price is_________

2. $\left(3^2\right)^2 =$

3. Find c in this right triangle. _____________

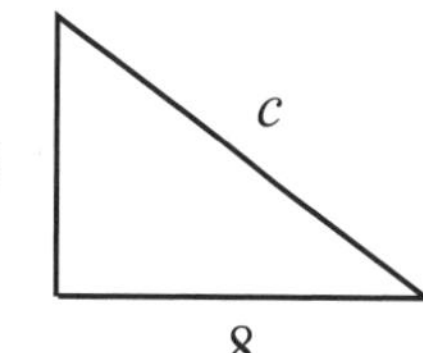

4. Complete this chart.

Sum	Product	Numbers
8	12	2 and 6
		3 and 8

5. If two angles in a triangle are 60° and 30°, is the third angle acute, obtuse, or right? _____________

6. $-3(14 + 3 \cdot (-4)) =$

7. Draw a line on this isosceles triangle to make two right triangles.

8. If the letter **A** is rotated 90° clockwise, what will it look like? _____________

9. What number is the arrow pointing toward? _________

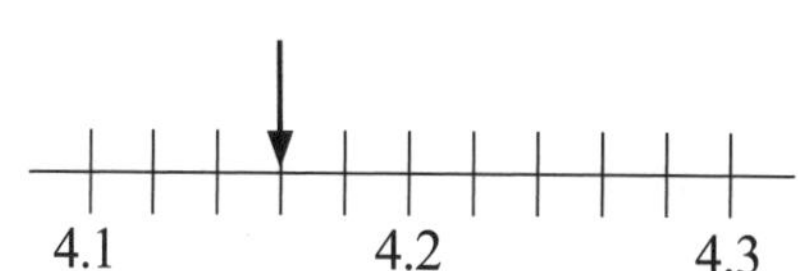

10. $\dfrac{\frac{6}{2}}{\frac{1}{3}} =$

MINUTE 97

1. Circle the fraction that is greater than $\frac{3}{14}$.

    a. $\frac{3}{20}$ **b.** $\frac{3}{15}$ **c.** $\frac{1}{4}$ **d.** $\frac{1}{7}$

2. Circle the measurement that is greater than 1 yard.

    a. 1 foot   **b.** 13 inches   **c.** 5 feet   **d.** 2 feet

3. Circle the amount that is greater than 0.06.

    a. $0.06\overline{1}$ **b.** 0.006   **c.** $\frac{1}{1,000}$ **d.** 4%

4. Circle the shape with more than nine sides.

    a. pentagon   **b.** hexagon   **c.** octagon   **d.** decagon

For Problems 5–6, use the figure to the right.

5. What percent of the squares have a black dot in them? __________

6. How many more black dots should be added so that $\frac{2}{3}$ of the squares would be filled? __________

7. A garden hose will be filling these boxes with water. Which box will take longer to fill? __________

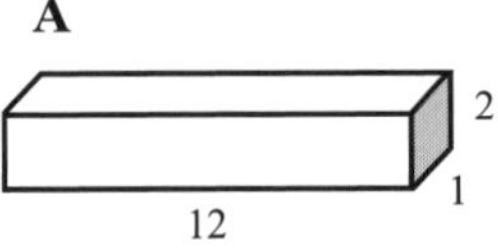

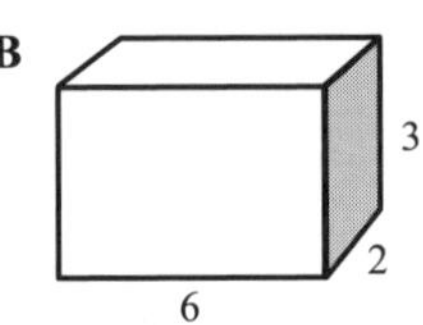

8. Fill in the missing numbers to complete the pattern.

1.5	3	6			48

9. $2\boxed{} \times 8 = 208$

10. $\sqrt{\sqrt{81}} =$

MINUTE 98

1. On Saturday, Justin drove 52 miles per hour for three hours. How far did he go?

____

2. Atlanta's population is two million, eight hundred thirty-three thousand, five hundred eleven. Write this number in standard form (using numbers). _______________

For Problems 3–5, use the map and chart to the right.

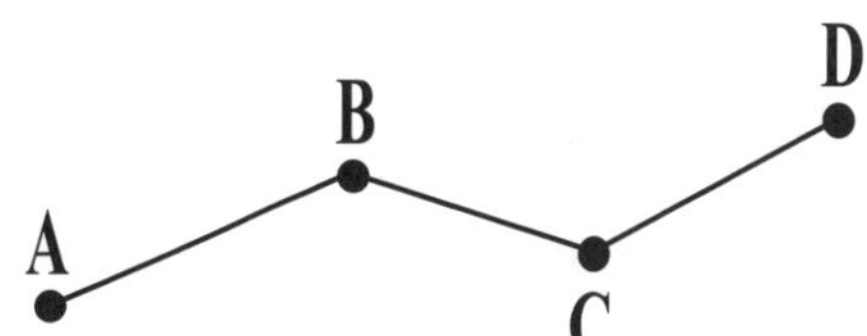

3. What is the road distance between towns A and C? _______________

4. What is the road distance between towns B and D? _______________

From	To	Distance
A	B	12 miles
B	C	7 miles
C	D	9 miles

5. If Marie rides her bike at a rate of seven miles per hour, how long will it take her to get from town A to town D? _______________

For Problems 6–10, match each statement with its correct answer.

6. $-4 \cdot 36$ **a.** -84

7. square root of 121   **b.** 144

8. 15% of 60   **c.** 11

9. $-42 \div 0.5$ **d.** 9

10. 12^2 **e.** -144

Minute 99

For Problems 1–3, use the chart to the right.

Bob's Camping Rental Store

Camping Supplies	Price Per Day
Mountain Bikes	$25
Climbing Gear	$15
Tents	$20
Canoes	$30
Backpacks	$10

1. Beth needs to rent a bike and tent for two days. How much will this cost her? _______________

2. Bryce needs to rent a backpack, canoe, and tent for three days. How much will this cost him?

____

3. Bob will offer a 10% discount if you rent an item for five or more days. How much would a tent cost to rent for five days? _______________

4. Circle the numbers that are composite.

10   16   21   23   25   29   30

5. Circle the number that is NOT divisible by 6.

12   15   18   24   30   36   48

For Problems 6–10, circle *Always true*, *Sometimes true*, or *Never true*.

6. The radius of a circle is half the diameter.

Always true   Sometimes true   Never true

7. A negative plus a positive is a negative.

Always true   Sometimes true   Never true

8. The diameter of a circle passes through the center of a circle.

Always true   Sometimes true   Never true

9. A negative times a negative is a negative.

Always true   Sometimes true   Never true

10. The perimeter of a shape is more than its area.

Always true   Sometimes true   Never true

MINUTE 100

For Problems 1–4, circle *True* or *False*.

1. $\dfrac{4}{5} = \dfrac{12}{16}$ True or False

2. If two triangles are similar, their sides are the same length. True or False

3. If the rate is consistent, 9 miles in 10 minutes = 4.5 miles in 5 minutes. True or False

4. If $\dfrac{x}{15} = \dfrac{9}{45}$, then $x = 3$. True or False

For Problems 5–8, circle the correct measurement.

5. The platform diving board is 33 (feet, inches, miles) high.

6. In tennis, it is possible for the ball to travel over 100 (miles, inches, feet) per hour.

7. In gymnastics, the balance beam is only 4 (inches, feet, yards) wide.

8. A softball weighs just under 7 (ounces, pounds, tons).

9. A hose will fill these boxes with water. Which box will take longer to fill? _________

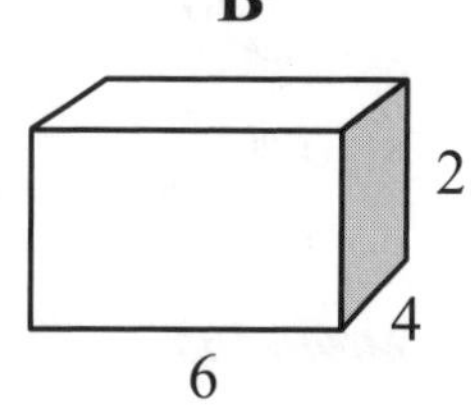

10. What percent of the squares have a black dot in them? _______

Minute Answer Key

Minute 1
1. 120
2. 21/100
3. 0.4, 40%
4. 1/2
5. 
6. 25 cm
7. 2
8. 14
9. 9
10. a

Minute 2
1. 2
2. >
3. 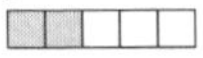
4. d
5. 3
6. 10
7. Scott, Annie
8. 24
9. 5
10. 36

Minute 3
1. 12
2. 1/12
3. 62%
4. •, +
5. 7
6. b
7. 27
8. =
9. >
10. >

Minute 4
1. 5.6
2. 57.6
3. 5
4. 5
5. → 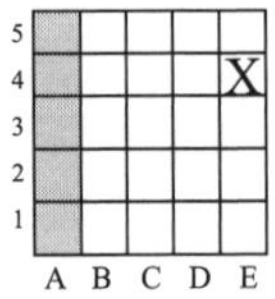

6. a
7. d
8. c
9. d
10. b

Minute 5
1. 0.45
2. 16
3. 12/50
4. 3, 5, 11, 17, 19
5. → 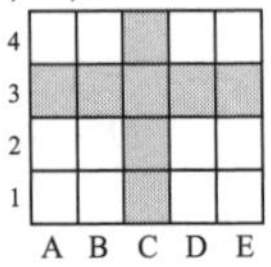

6. C3
7. a
8. 4
9. 3
10. 4

Minute 6
1. 1.6
2. 9
3. 2, 9
4. 8
5. 7
6. 4
7. False
8. True
9. True
10. True

Minute 7
1. 0.36
2. 4
3. Greatest: 78/100
   Least: 50%
4. 21, 14, 35
5. b
6. 36 cm
7. less than
8. 3
9. 8
10. 6

Minute 8
1. $4^2, 5^2, 6^2$
2. b
3. 1/4
4. 2/8, 3/12
5. c
6. 6
7. No
8. 15
9. 30
10. 4

Minute 9
1. 4 + 3 • 5
2. 7/12
3. 30
4. 190
5. Tom, Kyle
6. True
7. 37
8. $520
9. 8→16; 5→10; 12→36
10. 8, 12, 16, 36

Minute 10
1. False
2. True
3. False
4. 12/2, 8/8, 2^2
5. 3/8
6. → 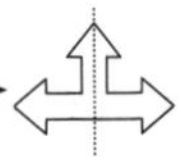
7. a
8. 6
9. 1/6
10. 1/3

Minute 11
1. 5, 3
2. 20
3. C3
4. 10 squares
5. 27
6. 3
7. 3.17
8. 1,001.5
9. 20
10. 7

Minute 12
1. 3/4
2. 2
3. False
4. 1/3
5. c
6. b
7. → 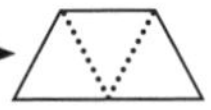
8. a
9. 17
10. 8

Minute 13
1. 9
2. 2.05
3. $0.\overline{912}$
4. d
5. 31, 37, 43
6. Yes
7. February
8. c
9. b
10. a

Minute 14
1. 3
2. 9
3. 14, 35, 42
4. 7, 3
5. 572, 527, 752, 725
6. d
7. e
8. a
9. b
10. c

Minute 15
1. 12
2. 1
3. 7
4. (4 + 5) • 2 = 18
5. True
6. a = 4, b = 100
7. d
8. Shade: Triangle
   Cross out: Hexagon
9. 1:00
10. c

Minute 16
1. Greatest: 3.3
   Least: 0.3
2. 32
3. b
4. 1/4
5. 2 and 8
6. 
7. 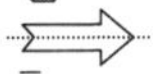
8. =
9. >
10. <

Minute 17
1. a
2. d
3. 7 3/4
4. 57.6
5. a
6. 15 units
7. 3/5
8. 60%
9. 60
10. 40%

Minute 18
1. 7/10
2. Monday
3. Saturday, Sunday
4. No
5. 50%
6. 
7. 4 and 8
8. d
9. c
10. 3

Minute 19
1. 0.97
2. 3.283
3. 180 pages
4. 8
5. Any five squares can be shaded.
6. 3
7. 42
8. 1
9. 14
10. 10

Minute 20
1. 3
2. 26
3. 6
4. 7.7
5. shade 6 more squares
6. 40
7. 2→4, 4→16, 6→36, 7→49
8. A
9. True
10. 2,500

MINUTE ANSWER KEY

MINUTE 21
1. False
2. False
3. True
4. a and c
5.
6. 4
7. girl birthdays
8. 14
9. 14
10. **A**

MINUTE 22
1. 16
2. -4, -1, 4
3. -7, 0, 8, 10
4. 3/4
5. 4 and 10
6. c
7. <
8. =
9. >
10. =

MINUTE 23
1. see chart
2. see chart
3. see chart
4. 56
5. see circle
6. 12
7.
8. d
9. 5
10. 20

MINUTE 24
1. 36
2. 36
3. 0.038
4. Sunday
5. Circle: 23rd
6. Put an X: 1, 4, 9, 16, 25
7. Shade: March 11
8. 2,600
9. True
10. c

MINUTE 25
1. 10, 10
2. 4
3. 25
4. Shade: 7, 14, 21, 28, 35
5. Circle: 13
6. 34
7. $5.50
8. 5
9. d
10. b

MINUTE 26
1. 5/11
2. c
3. 0.13467
4. 12.4
5. 7, 2, 2
6. 28
7. Serena
8. b
9. d
10. True

MINUTE 27
1. 6/25
2. 1/4
3. Circle: 4 Box: 15
4. 9 in. × 6 in.
5. False
6. 32
7. a
8. <
9. >
10. >

MINUTE 28
1. 52
2. 22
3. 10.2 cm
4. 3^4, 3 • 3 • 3 • 3
5. xy, $x(y)$, $(x)(y)$, $x • y$
6. True
7. Ten donuts for $2
8. c
9.
10. Each shape has one more side.

MINUTE 29
1. 0.85
2. 0.06
3. 26.8 cm
4. 10
5. d
6. c
7. e
8. d
9. a
10. b

MINUTE 30
1. Ray
2. a
3. $3 + (6 - 2) • 4 = 19$
4. Yes
5.

MINUTE 31
6. 2, 3, 4
7. 3/2
8. 5,280 feet
9. 2,000 pounds
10. 4 quarts

MINUTE 31
1. 5, 1
2. 7
3. 4
4. 8 sq. units
5. 12 units
6. 32%
7. 5, 10, 15
8. 1:15
9. 9,961
10. 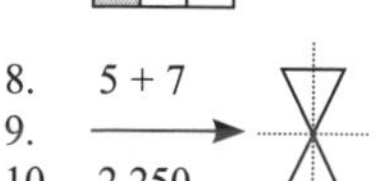

MINUTE 32
1. →
2. No
3. 19th
4. $310
5. 9
6. c
7. 60
8. grams
9. meters
10. centimeters

MINUTE 33
1. 35, 48
2. $2.10
3. 19
4. 8
5. A, C, E
6. E
7. 11
8. 12
9.
10. 65%

MINUTE 34
1. 6,000
2. 18
3. 5 and 7
4. $\overline{XY}$
5. 9
6. →
7. 16/3
8. 7.4
9. 9/40
10. a gray square

MINUTE 35
1. $36.18
2. 9
3. 2/5

MINUTE 36
4. 3/5
5. c
6. $5m$
7. 14
8. 5
9. 1%
10. RED

MINUTE 36
1. 1 3/4
2. 1/5
3. 3/8
4. 7
5. 1,000
6. 28 units
7. 30 sq. units
8. hot dogs: 10, hamburgers: 8, both: 4
9. parallel
10. M12

MINUTE 37
1. 2
2. 5/6
3. 0.51
4. 0.05
5. c
6. Multiples of 4: 4, 16, 20
   Multiples of 6: 18 Both: 12
7.
8. 5 + 7
9.
10. 2,250

MINUTE 38
1. 1/10
2.
3. 1.8
4. c
5. 0.023
6. 45.7
7. 0.05
8. length × width × height
9. 20
10. red

MINUTE 39
1. 1
2. 38.717
3. 12
4. 27
5. 32
6. V5
7. c
8. d
9. a
10. b

MINUTE ANSWER KEY

MINUTE 40
1. 108, 130
2. 0.058, 0.085, 0.508, 0.580
3. 

4. 5.5
5. 95
6. 90
7. 15 people
8. 3 in.
9. 1,109, 10^4, $\sqrt{1\ billion}$
10. 0

MINUTE 41
1. 0.3, 0.33, 3.0, 3.3
2. 2, 6, 15
3. Saturday
4. Thursday
5. 3
6. Multiples of 5: 5, 20, 30
   Multiples of 7: 14, 21
   Both: 35
7. 35
8. 3
9. 12
10. 64

MINUTE 42
1. No
2. c
3. 3/10
4. golf
5. baseball, basketball
6. baseball, football
7. 30
8. 0
9. No
10. Yes

MINUTE 43
1. Shade 4 additional squares (6 total)
2. 15.99
3. 2.222
4. 500
5. 85
6. d
7. 85
8. 13 people
9. d
10. 2.6583

MINUTE 44
1. 1/12
2. 3/5
3. 2/5
4. Saturday
5. No
6. 300 cm, 0.6 m
7. 3,200 g, 0.06 kg
8. 108 in., 4 yards
9. 4.6
10. e (no right angle)

MINUTE 45
1. 11.64
2. 0.68
3. 1.2
4. perpendicular
5. 125
6. 451 (combination of odd and even numbers)
7. True
8. False
9. False
10. True

MINUTE 46
1. 81
2. 34.67
3. 8, 9, 10, 14
4. d
5. $18
6. 
7. 4
8. 941, 914
9. 8, 10
10. 

MINUTE 47
1. 2 hours, 12 minutes
2. 30%
3. B, D or A, C
4. Passed
5. 7/12
6. c
7. b
8. d
9. a
10. e

MINUTE 48
1. 5
2. 26 units
3. 20 sq. units
4. 12
5. 52
6. d
7. a
8. e
9. c
10. b

MINUTE 49
1. 4.02
2. 3/4
3. 0.25
4. 12.8 units
5. 8.8 sq. units
6. No
7. 128
8. d
9. a
10. b

MINUTE 50
1. 6 squares shaded
2. 85%
3. 26 units
4. → 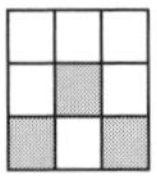
5. 7
6. 7
7. 7
8. 6
9. 15
10. 36

MINUTE 51
1. 56
2. -42
3. negative
4. 25
5. 0.5
6. 5, 17
7. 3,426
8. 300 units^3
9. 7/10
10. 3/10

MINUTE 52
1. -5
2. -13
3. 64
4. 3
5. 9 blocks
6. -3
7. -60
8. -12
9. 30
10. 2

MINUTE 53
1. -5
2. 48
3. 6
4. 2
5. 3
6. D
7. B, C
8. A, B
9. a
10. No

MINUTE 54
1. 10
2. -3
3. 4
4. 1
5. 101
6. 6
7. 3
8. 2
9. -10, -5, 0, 5, 10
10. $3^2, (-2)^2, 0, -5$

MINUTE 55
1. 6
2. → 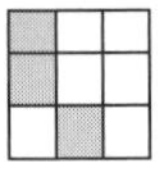
3. d
4. c
5. D
6. (3, -3)
7. (-4, 3)
8. d
9. >
10. <

MINUTE 56
1. +, ÷
2. 5
3. 2
4. $0.0\overline{12}$
5. 6
6. 4 3/4
7. False
8. False
9. True
10. True

MINUTE 57
1. 14
2. 4
3. 3
4. goes up
5. 3
6. -4
7. J12
8. 2
9. 22
10. TH, TT

MINUTE 58
1. –, +
2. -27
3. 80°
4. 8
5. c
6. b
7. b
8. 5
9. -10
10. -6

MINUTE 59
1. 125°
2. 2.03, 2.22
3. 2
4. c
5. c
6. 3 sides, then 4 sides
7. 12
8. 1/8 of the bag
9. -5
10. -4

MINUTE ANSWER KEY

MINUTE 60
1. b
2. 12,489
3. Yes
4. 59
5. c
6. 6.6, 7.2
7. -6, 5, 8
8. Tuesday
9. Saturday
10. c

MINUTE 61
1. 150 cm^2
2. 0.6
3. $4 + 6 + 8 + 9 = 27$
4. 8
5. 1
6. c
7. 27
8. 26.789
9. E25
10. 140°

MINUTE 62
1. 3/4
2. C
3. D
4. A
5. 2
6. -34
7. $40
8. -5
9. 7 (prime)
10. 1 3 5 7 9 7 5 3 1

MINUTE 63
1. b
2. 24
3. 41
4. 1
5. 2
6. down
7. 7, 3
8. 20
9. 60 m^2
10. -2

MINUTE 64
1. c
2. d
3. e
4. b
5. a
6. 8
7. 5, 6, 7
8. 2/7
9. 21/22
10. 2, -1, 6

MINUTE 65
1. -28, -48
2. $2n + 1 = 11$
3. 5
4. 5 (prime)
5. 7/12 (doesn't reduce to 1/2)
6. (not acute)
7. b
8. c
9. a
10. d

MINUTE 66
1. C
2. D
3. b
4. a
5. False
6. 20 people
7. 1/5, 20%
8. >
9. =
10. <

MINUTE 67
1. 1/8
2. 6
3. -6, -6
4. 14
5. a
6. 3
7. a
8. 8:00
9. 10 miles
10. 2:00, 6:00

MINUTE 68
1. 14, 21, 42
2. 8
3. 11
4. 9, 12, 15
5. Quadrant II
6. (-3, 2)
7. Quadrant IV
8. D4
9. A1
10.

MINUTE 69
1. -3, 3
2. 60
3. a
4. bed, dining table
5. paper clip, pencil eraser, bottle cap
6. 24 in.^{3}
7. c
8. b
9. d
10. a

MINUTE 70
1. Numbers in A get doubled in B.
2. 1/16
3. c
4. b
5. c
6.
7. 75%
8. Quadrant I
9. Quadrant III
10. positive slope

MINUTE 71
1. 55
2. No
3. 20
4. Red
5. Blue
6. 30
7. True
8. 3
9. $y = 3x$
10. -9

MINUTE 72
1. 3 times
2. 1 number
3. 4.1
4. -9, -13
5. c
6. a
7. quotient
8. difference
9. product
10. sum

MINUTE 73
1. 5, 3
2. +, ÷
3. - 0.7
4. Yes
5. -18, -40
6. 31.4
7. d
8. c
9. a
10. b

MINUTE 74
1.
2. 35
3. 48
4. 24
5. 56
6. 3:30
7. median
8. acute
9. mode
10. range

MINUTE 75
1. 4/5
2. 2,400
3. -8
4. 60
5. -11, -11
6. 42
7. c
8. C, 3
9. B
10. 19

MINUTE 76
1. 22
2. $0.60
3. 70 degrees
4. $15
5. $8.20
6. 205
7. 105
8. 164
9. 19, 2
10. 7.5 cm

MINUTE 77
1. 20%
2. 125
3. -8
4. 21
5. 28 in.
6. No
7. c
8. d
9. b
10. a

MINUTE 78
1. A
2. B
3. C
4. $2.50
5. 40
6. $10
7. -16
8. 60
9. Quadrant IV
10. 4

MINUTE 79
1. 3
2. 3, 5
3. 4
4. $250
5. 1, -11
6. 10 cubes
7. 60
8. d
9. b
10. a

MINUTE ANSWER KEY

MINUTE 80
1. 74 minutes
2. 1.5 feet
3. 3^3, 27, 3 • 3 • 3
4. 100, $\frac{1,000}{10}$, 10^2
5. ⬠, Pentagon, ⬠
6. 1.265×10^2, 0.1265×10^3, 12.65×10^1
7. 3.2 miles
8. 28 legs
9. 25, 36
10. c

MINUTE 81
1. 0.85 or 17/20
2. 86.521
3. >
4. <
5. >
6. Parallel
7. $45 \div (-5) = -9$
8. $-4 \cdot 2 = -8$
9. $-9 \div 2 \cdot 6 = -27$
10. $-8 + -7 = -15$

MINUTE 82
1. 10
2. perpendicular
3. (-3, 4)
4. Quadrant II
5. Quadrant III
6. $y = x + 3$
7. c
8. 36 inches
9. 180
10. 360

MINUTE 83
1. 48 ft.
2. →
3. True
4. False
5. True
6. True
7. 0.68
8. 3
9. 6
10. 24

MINUTE 84
1. 5
2. 6
3. 52
4. 1
5. 6 days
6. 0.62
7. volume
8. area
9. perimeter
10. prime

MINUTE 85
1. c
2. 1/13
3. 7 1/4 or 29/4
4. 29
5. three pentagons
6. 11
7. c
8. 12
9. 3
10. 64

MINUTE 86
1. 132 minutes
2. 6
3. 12
4. 25
5. 5
6. 144
7. Quadrant II
8. -8
9. -4
10. 2

MINUTE 87
1. -5
2. -24
3. 0
4. 0.87
5. 10
6. c
7. parallel
8. 0, 2
9. IV
10. -3

MINUTE 88
1. a
2. 25%
3. 5^2 or 25
4. a
5. 14
6. 3
7. a^4
8. 4/25
9. 4
10. 30%

MINUTE 89
1. $3^4 \cdot 5^2$
2. b
3. 50
4. 5.5
5. c
6. 12
7. 16
8. d
9. 1 person
10. B, C

MINUTE 90
1. Lost $1
2. 21
3. 28%

MINUTE 91
1. b
2. 15
3. 2/5
4. 40%
5. 0.4
6. mean
7. 23/7
8. 27
9. 2, 3, 5
10. 60°

MINUTE 92
1. 12
2. 30, -28
3. 29/2
4. a
5. 10%
6. 50
7. 2
8. =
9. >
10. <

MINUTE 93
1. 9
2. -3
3. -3
4. -7, -5, 0, |−5|
5. 4^{11}
6. acute
7. 150
8. (-3, 4)
9. III
10. Yes

MINUTE 94
1. 13
2. $2a + 2b$
3. ab
4. 34
5. 8, 12, 20
6. d
7. e
8. c
9. a
10. b

MINUTE 95
1. $6.40
2. 0.57
3. -2/3
4. 24, -56, -36
5. 2
6. $4a$
7. 49 sq. units

4. second
5. -15/56
6. a, b, c
7. 4^5
8. 15
9. 26
10. 6.5 inches

MINUTE 96
1. 20%
2. 3^4 or 81
3. 10
4. 11, 24
5. right
6. -6
7. →
8. A
9. 4.16
10. 9

MINUTE 97
1. c
2. c
3. a
4. d
5. 33 1/3%
6. 5 dots
7. B
8. 12, 24
9. 6
10. 3

MINUTE 98
1. 156 miles
2. 2,833,511
3. 19 miles
4. 16 miles
5. 4 hours
6. e
7. c
8. d
9. a
10. b

MINUTE 99
1. $90
2. $180
3. $90
4. 10, 16, 21, 25, 30
5. 15
6. Always true
7. Sometimes true
8. Always true
9. Never true
10. Sometimes true

MINUTE 100
1. False
2. False
3. True
4. True
5. feet
6. miles
7. inches
8. ounces
9. Box A
10. 30%

8. 6
9. -1, 20, -7
10. 13, 25, 36, 38